*Cambridge Imperial and Post-Colonial Studies Series*

General Editors: **Megan Vaughan**, Kings' College, Cambridge, and **Richard Drayton**, Corpus Christi College, Cambridge.

This informative series covers the broad span of modern imperial history while also exploring the recent developments in former colonial states where residues of empire can still be found. The books provide in-depth examinations of empires as competing and complementary power structures encouraging the reader to reconsider their understanding of international and world history during recent centuries.

Iftekhar Iqbal
THE BENGAL DELTA
Ecology, State and Social Change, 1843–1943

Brian Ireland
THE US MILITARY IN HAWAI'I
Colonialism, Memory and Resistance

Robin Jeffrey
POLITICS, WOMEN AND WELL-BEING
How Kerala became a 'Model'

Gerold Krozewski
MONEY AND THE END OF EMPIRE
British International Economic Policy and the Colonies, 1947–58

Javed Majeed
AUTOBIOGRAPHY, TRAVEL AND POST-NATIONAL IDENTITY

Francine McKenzie
REDEFINING THE BONDS OF COMMONWEALTH 1939–1948
The Politics of Preference

Gabriel Paquette
ENLIGHTENMENT, GOVERNANCE AND REFORM IN SPAIN AND ITS EMPIRE 1759–1808

Sandhya L. Polu
PERCEPTION OF RISK
Policy-Making on Infectious Disease in India, 1892–1940

Jennifer Regan-Lefebvre
IRISH AND INDIAN
The Cosmopolitan Politics of Alfred Webb

Ricardo Roque
HEADHUNTING AND COLONIALISM
Anthropology and the Circulation of Human Skulls in the Portuguese Empire, 1870–1930

Jonathan Saha
LAW, DISORDER AND THE COLONIAL STATE
Corruption in Burma c.1900

Michael Silvestri
IRELAND AND INDIA
Nationalism, Empire and Memory

John Singleton and Paul Robertson
ECONOMIC RELATIONS BETWEEN BRITAIN AND AUSTRALASIA 1945–1970

Julia Tischler
LIGHT AND POWER FOR A MULTIRACIAL NATION
The Kariba Dam Scheme in the Central African Federation

Aparna Vaidik
IMPERIAL ANDAMANS
Colonial Encounter and Island History

Jon E. Wilson
THE DOMINATION OF STRANGERS
Modern Governance in Eastern India, 1780–1835

Cambridge Imperial and Post-Colonial Studies Series
Series Standing Order ISBN 978–0–333–91908–8 (Hardback)
978–0–333–91909–5 (Paperback)
(outside North America only)

You can receive future titles in this series as they are published by placing a standing order. Please contact your bookseller or, in case of difficulty, write to us at the address below with your name and address, the title of the series and the ISBN quoted above.

Customer Services Department, Macmillan Distribution Ltd, Houndmills, Basingstoke, Hampshire RG21 6XS, England

# Ayurveda Made Modern

## Political Histories of Indigenous Medicine in North India, 1900–1955

Rachel Berger

*Associate Professor, Department of History, Concordia University, Canada*

palgrave
macmillan

First published 2013 by
PALGRAVE MACMILLAN

Palgrave Macmillan in the UK is an imprint of Macmillan Publishers Limited, registered in England, company number 785998, of Houndmills, Basingstoke, Hampshire RG21 6XS.

Palgrave Macmillan in the US is a division of St Martin's Press LLC, 175 Fifth Avenue, New York, NY 10010.

Palgrave Macmillan is the global academic imprint of the above companies and has companies and representatives throughout the world.

Palgrave® and Macmillan® are registered trademarks in the United States, the United Kingdom, Europe and other countries.

ISBN 978-1-349-32968-7     ISBN 978-1-137-31590-8 (eBook)
DOI 10.1057/9781137315908

This book is printed on paper suitable for recycling and made from fully managed and sustained forest sources. Logging, pulping and manufacturing processes are expected to conform to the environmental regulations of the country of origin.

A catalogue record for this book is available from the British Library.

A catalog record for this book is available from the Library of Congress.

*For Nitika*

# Contents

# Preface and Acknowledgements

I have found, over the years, that studying Ayurveda invites enquiries as to the nature of my connection to the subject matter, fixed on ideological and spiritual axes. The assumption is that one has come to Ayurveda through an experience of practice, either as practitioner or lay follower, and has a personal commitment to the field. I have disappointed enquirers for over ten years with my claim to a simply academic commitment to Ayurveda – and to its politics and its challenges to governance, no less – rather than the intricate machinations of its internal logics. I am most acutely aware of the ways in which this book pays little notice to these substantial changes and instead uses Ayurveda to ask other questions.

I came to this project through a deeper engagement with the politics and experiences of women's health in colonial and postcolonial India. I spent the last days of my undergraduate degree working as Research Assistant to the formidable scholar and activist Shree Mulay, who introduced me to the ways in which gender was deployed and entangled in the ongoing saga of development in the subcontinent. I marvelled at the activist struggle that she and her comrades were involved in to disentangle the rights of women – and all subject citizens – from the neoliberal experiments that violated their bodies and their autonomy. Searching for deeper answers about the contemporary context of Indian health politics, I spent my early years of graduate study uncovering the broader themes that connected gender and sexuality to ideologies of health, first in the secondary literature and later on in Hindi primary source material. I was well steeped in early twentieth-century recountings of gynaecology, obstetrics and birth control, in juxtaposition with readings of critical theory that engaged postcolonial and poststructuralist approaches to the body in its colonial context.

A funny thing happened, though, as I poured through those health manuals and identified the shape and form of popular discourse. I kept running into these markers of a medical past, sometimes deployed in linguistic turns of phrase (like the ever-present *pracheen samay mei*, in olden times), sometimes referred to affectionately as 'our mother's practices', but consistently lying just beneath the modern texts in the limbo of passed time. I found it both impossible to identify this deferred set of practices, while equally unsatisfied just moving beyond it, as other colleagues rightly have. In essence, the study that has become this manuscript was borne out of that moment of curiosity: of simply trying to piece together an imagined background and to figure out the trajectory of deferred practice in modern times.

I started looking for Ayurveda in the places carved out by established scholars who worked on the indigenous medical systems. The beginning and end points for this project were always the lay conversations about Ayurveda that I've explored in Chapter 3, and I was sure that the scope of fieldwork and dissertating would be taken up with the work of tracking texts, translating them and constructing a social history extrapolated from the dynamic space of popular discourse in North India. Trips to the official archives of the Government of India in both London and India were meant to be a formality, where I would find very little, perhaps enough to get a Government of India perspective on the indigenous medical systems which I assumed would be dismissive and brief.

In Delhi, this was the case, entirely due to the logic of colonial governance and the bureaucratic histories housed there. Confined to time and space before 1914, the documents that were categorized (and that I could easily call up – most carried that mark of having been 'not transferred') were almost entirely dismissive of indigenous medicine, or, in rare, instances, brought questions of their value to the floors of disinterested political assemblies. With a couple of exceptions – which we will explore in Chapters 1 and 2 – the files behaved as I thought they would and I was dedicated once more to social history through cultural research.

A chance trip to Lucknow, however, shook up the project and my thinking about the trajectories of health in modern India. A brief glance one morning through the guides to records revealed a huge engagement with indigenous medicine, even out of proportion with other concerns of health governance. A short trip turned into a long one as the bureaucratic history of the Indigenous Systems of Medicine unfolded in front of me that dated from the early 1920s well into the postcolonial period, becoming, shockingly, more voluminously documented in the early 1950s than at any other time. Working backwards from these files, I pieced together a bureaucratic history through diarchic devolution and, eventually, through Congress governance that ceased to leave Ayurveda in the corner – or in the ancient past. Instead, it centred indigenous medicine in very broad, very upper-level conversations about the structures of governance in the province, and, ultimately, within the imaginings of an independent republic.

The foundation of the final iteration project is in this discovery: that Ayurveda had a political history, that it posed a problem for the state, that it was both an object and subject of governance. Moreover, this history had a big hand in making it modern – beyond the revamping of the internal substance of the system, or the formalization of practitioners into guilds or associations, or the introduction (or conscious dismissal) of allopathic techniques and technologies. Ayurveda was not only extant on the sidelines – it was discussed, in complex terms and with a variety of outcomes, at the centre of power. Moreover, it was deployed in very familiar ways within an emergent biopolitic of late colonial rule.

The study took form as I rerouted questions of power and governance that historians have often asked about health through trajectories that were less familiar and predictable than the usual ones engaged when writing about health politics in colonial times. Rather than a top-down approach, I wanted to write a history of health politics that began outside of the spheres of biopolitical engagement and that could chart the ways in which Ayurveda worked its way to the centre. I have framed this book as a study of the ways in which Ayurveda became governable. I have suggested a reading of the entrance of tradition into modernity that is staked around politics, and not around some of the more familiar lenses we have for envisioning that phenomenon. Rather than thinking about what is lost or gained, or dwelling heavily on the politics of identity and authenticity (though they are certainly there in Chapters 4–6), I instead wanted to tell a story that moved beyond affect and that thought through the ramifications of this move. In the logic of colonialism, Ayurveda should never have been privy to formal politics. That it would become a concern of the state – and that it would become a way in which a new state defines its responsibilities – seemed light years away from my initial hypotheses about its exclusion from the realm of the formal politics. That the work it ultimately does returns us to the realm of the social through the enactment of communal violence (in both acute and banal ways) is another part of its legacy, tying it in once more to the nationalist and identitarian inflections of the health guides that helped initiate the project.

I hope the result is an Ayurveda turned on its head. The brilliant work of thinking about Ayurveda from the inside out is not the terrain of this book. Instead, I have tried to use Ayurveda to think about the interwar and postcolonial state of politics from the perspective of a tenuous, seemingly dismissible, suspiciously 'unscientific' system that posed a problem for governance, and a challenge to the structures that be. In essence, to look at the insides of the this thing we call the state from the perspective of a recalcitrant outsider that is tamed into being governed but still remains somewhat aloof. As such, I posit an Ayurveda that is anything other than amorphous, ahistorical and apolitical – an Ayurveda that is in motion, made modern through its governability, made relevant through the ways in which it could link the political meaningfully to the realm of the social.

In pursuit of this variety of Ayurvedic history, I have acquired many debts that I am happy to have the opportunity to acknowledge.

A Commonwealth Scholarship and a doctoral fellowship from the Social Science and Humanities Research Council of Canada funded my years of study at the University of Cambridge. A Shastri Indo-Canadian Institute doctoral fellowship along with funds from the Prince Consort Fund and the Clare College Travel Fund provided funding for my research trips to India. Grants from the Fonds de Recherche sur la Société et la Culture (FQRSC) and from Concordia University funded later research trips and the eventual compilation of the manuscript. The research for this project was conducted at Teen Murti, the National Archives of India, the Marwari Library in Delhi, Kala Bhavan in Banaras, the medical and Ayurveda faculty libraries at Banaras Hindu University (BHU), the Uttar Pradesh State Archive, the Allahabad and Delhi Municipal Archives, the Oriental and Indian Office Collection, the British Library, the Bodleian, the South Asia Institute at Oxford, the Centre for South Asian Studies, SOAS library, Cambridge University Library, the Wellcome Library for the History of Medicine, and the Osler Medical Library and Birks Religious Studies Library at McGill University. The staff at the Cambridge Centre for South Asian Studies provided ongoing help and support. Kamala Chaturvedi at BHU medical library went out of her way to help me on too many occasions. I'd like to thank Bob, Kwame and Emma at the Oriental and India Office Collection (OIOC) for lobbying hard to get me my Hindi sources quickly (and for their general good humour). The Misra brothers at the Marwari Pustakalya in Delhi were exceptionally helpful, and I thank them for their hospitality, friendship and help.

Excerpts of the text appeared previously in 'From the Biomoral to the Biopolitical: Ayurveda's Political Histories' in *South Asia History and Culture*, 4, 1 January 2013: 48–64, and 'Between Digestion and Desire: Genealogies of Food in Nationalist North India' *Modern Asian Studies*, Cambridge University Press, Firstview 2 (2013): 1–22. I would like to thank these journals for permission to reprint this material.

This project began at the University of Toronto, where Michelle Murphy pushed me to pursue the questions about health, the body and the state that I was only just beginning to formulate. At Cambridge, Polly O'Hanlon, my PhD supervisor, took great pains to balance the histories of language, politics, culture and medicine, and guided me steadily beyond the chronological and conceptual barriers of the modern era. I am greatly indebted to her for opening methodological and substantive doors, and for her admirable patience and support of my project. Francesca Orsini taught me about the importance of language and cultural production to the formation of politics. Most of all, she deserves my thanks for teaching me Hindi and for the hours of slogging through obscure medical texts without squirming. Aishwarj Kumar, Virendra Singh and Sandesh Singh were wonderful and patient Hindi teachers. I am immensely grateful for Sarah Hodges' guidance and support from the earliest days of the project until the final days of its conclusion; she remains an intellectual and professional inspiration. I would also like to thank Richard Drayton and Megan Vaughan for including the book in the *Cambridge Studies in Imperial and Postcolonial History* series. Holly Tyler and Jen McCall provided wonderful editorial support and great patience.

This project is indebted to the generous colleagues and mentors with whom I've discussed its objectives over the years. Many thanks to Seema Alavi, Chris Bayly, Ritu Birla, the late Raj Chandravarkar, Markus Daechshel, Michael Dodson, Richard Drayton, Waltraud Ernst, Doug Haynes, Stephen Legg, Laurence Monnais, Barbara Ramusack, Peter Robb, Tanika Sarkar, Katherine Butler Schofield, Sujit Sivasundaram, Fred Smith, Noemi Toussignant, Megan Vaughan and Dominik Wujastyk. Special thanks to Guy Attewell, Projit Mukharji and Ishita Pande, who grappled with similar questions during our time as students together and in subsequent years; I am especially grateful to have been in the same place at the same time as Ishita on several occasions. Dagmar Wujastyk helped me navigate the world of Ayurveda-inflected Indology. Similarly, the anonymous reviewer for Palgrave Macmillan provided commentary that greatly improved the final product.

I join a long list of scholars in thanking Ram Advani for his hospitality, wisdom and endless cups of *chai* shared at his shop in Lucknow; I am also grateful to the late Mrs Sharma for her hospitality in Lucknow. Special thanks go to Ramesh Chetwal and Ammaji, and Arun and Simmi Khanna, who have all been excellent hosts in Delhi.

Closer to home, Sarah Beth Hunt let me try out every idea on her, and read multiple versions of the manuscript in its various stages. Jacob

Copeman has the greatest propensity for *daan*. Mitra Sharafi's generosity has been an inspiration. Justin Jones and Ben Hopkins were the best of company in Delhi, Lucknow and Cambridge. This project was sustained by years of conversation across continents and cups of *chai* (and other substances) with Emma Alexander-Mudaliar, Sunil Amrith, Shahana Bhattacharya, Rohit De, Nigel Dixon, Mike Finn, Kaveri Gill, A.L. Gust, Suzanne Haselle-Newcombe, Ananya Jahanara Kabir, Steve Legg, Jean Meiring, Eleanor Newbigin, Rosie Peppin-Vaughan, Jahnavi Phalkey, Pedro Ramos-Pinto, Emma Reisz, Katherine Butler Schofield, Uditi Sen, Taylor Sherman, Julie Stephens, Abi Wills and Ben Zachariah. Charlotte Cooper, Xtina Lamb, Nikhil Chandra and Rahul Suchde provided homes away from home and unfailing hospitality over many trips to London.

It has been a privilege to end up at the place where my journey into South Asian history began. At Concordia University Norman Ingram, Shannon McSheffrey and Ron Rudin gave me the administrative support and professional freedom to do justice to the project. My fellow Fellows, along with Viviane Namaste and Gen Rail, have been wonderful interlocutors at the Simone de Beauvoir Institute. I am fortunate to work and live in the company of Dima Ayoub, Danielle Bobker, Meredith Evans, Kevin Gould, Andy Ivaska, Wilson Chacko Jacob, Prashant Keshavmurthy, Zohar Kfir, Pasha Ahmad Khan, Erica Lehrer, Katherine Lemons, Lisa Lynch, Ted McCormick, Khalid Medani, Leslie Orr, Monica Patterson, Sina Queyras, Elena Razlogova, Ted Rutland, Megha Sehdev, Gavin Taylor, Tim Sedo, Theresa Ventura, Tracy Zhang and Anya Zilberstein. Johanna Ransmeier came along late in the game but just in time. Jon Soske breathed some life back into this project at a crucial moment, for which I will be forever grateful. Megan Webster brought me back from the brink on several occasions. I owe lifetimes of gratitude to my feminist cheersquad: Deb Lunny, Gada Mahrouse, Tanisha Ramachandran, Trish Salah and Sarwat Viqar, who have brought home time and again the amazing potential of feminist collaboration. Finally, I am humbled by the friendship and support of Amanda Gauthier, Tanya Schuh, Anita Dhami, Gen Vallerand and Sherrie de la Paz.

Dolores Chew, John Hill and Karen Ray have been the most gracious teachers at every interval, and I am immeasurably lucky to have had the privilege of their guidance, support and conversation since my earliest student days. I hope this book has done justice to the nuances of the South Asian past that they hoped I would take on board.

My family has done the work of sustaining this project and its author through many ups and downs. My parents-in-law, Rajni Kent and Krishan Dosaj, have been incredibly encouraging and generous, both stimulating and gracious with their praise and interest. My siblings, Joey Berger, Vesna Antwan and Neel Dosaj, have been unfailingly good-humoured. Finally, my parents, Howard Berger and Ellayne Kaplan, took the project, and the long absences it required, on board with good cheer. While their interest in India grew after mine developed, the roots of the questions that frame this book come from their own engagements with community, social justice and the pursuit of knowledge. I thank everyone, officially, for picking up the many balls I dropped along the way.

Only Nitika Dosaj knows the debt that I, and this project, owe to her. She has turned a keen eye to the manuscript and challenged my thinking on the Indian past in unexpected ways. She has also kept the whole show running while one of its key players was on the road. Lucky for me, she doesn't keep tabs. This book is for her.

# Introduction: Ayurveda in Motion

Ayurveda is the oldest ongoing medical system in South Asia, with ancient Indian roots and global contemporary wings. Even the most mainstream international supermarkets contain boxes of Ayurvedic tea; spas and salons offer Ayurvedic massage; and major celebrities the world over advocate Ayurvedic diets and cleanses (characterized, mostly, by deprivation and the robust effects of fibre).[1] Associated with the most primal, basic iterations of a medical logic in the subcontinent, it has come to have an international career built upon the majestic possibilities of its mysterious Eastern depths. In India, too, the contemporary marketplace has revamped Ayurveda into a mode of selling luxury goods, with brands like Kama and Biotique offering pricy elixirs and cosmetics to middle-class (and upper middle-class) consumers wanting to buy into the affects of organic, indigenous health.[2] Moving beyond the realm of self-diagnosis (and cosmetic indulgence), the neighbourhoods of most Indian cities are papered with ads for Ayurvedic practitioners able to cure the whole gambit of modern illness, ranging from AIDS to diabetes to fertility (and everything in between).

Ayurveda, however, has yet to be historicized in modernity. Scholars have suggested that the lack of attention to the evolution of Ayurveda has resulted in the erroneous impression that the knowledge created at its inception was left untouched and pristine throughout the many centuries of its existence.[3] At best, its alleged dismissal from the realm of the biomedical has accounted for its exclusion from the scrutiny of political economy. Instead, Ayurveda is understood to be ancient in its logic, timeless in its applicability, preserved from the trappings of allopathic biomedicine and devoid of political significance. It is consistently located outside of the scope of biomedical logic – and, subsequently, biopolitical regimes. Across discourse, Ayurveda is employed as an alternative – at times nascent, at times explicitly aggressive – to the hegemonic nature of 'Western' medicine. The notion of the 'West' at work cuts both ways: on the one hand, Ayurveda is championed by consumers who cast its holistic approach as a positive departure from the rigid structures and micro-approaches of Western (or Allopathic) biomedicine; on the other, the same logic that casts Ayurveda as exception works just as easily to deny the system a permanent and serious seat at the ontological table of medical knowledge, innovation and investment. The focus on an authentic, indigenous past has separated Ayurveda from the larger processes of historical change at work in the subcontinent. Regardless, Ayurveda is implicated in the processes of creating historical meaning through its symbolic invocation of a past beyond history or memory.

This study attempts to reverse the representation of Ayurveda as a static placeholder by mapping its history in late colonial modernity within the evolving political regimes of the emerging Indian State. In it, I map out the biopolitics of the interwar and early postcolonial period in an effort to understand the ways in which the government of the United Provinces (UP) – and, later the Uttar Pradesh state government – conceptualized ideas of health, population and state responsibility in this period of massive political change. The study examines the shift from an entrenched colonial reticence to consider the Indigenous Medical Systems as legitimate scientific medicine to a growing acceptance of Ayurvedic medicine following the First World War. Locating the moment of transition within the implementation of a dyarchic system of governance in 1919, I argue that the revamping of the 'Medical Services' into an important new category of regional governance ushered in an era of health planning that considered curative and preventative medicine as key components of the 'health' of the population. This study will thus examine the ways in which the functions of the earlier medical infrastructure were retained, challenged or modified by this new agenda for health care. As such, it illuminates the way in which conceptions of power, authority and agency were newly configured and consolidated as politics were revamped in provincial India.

In so doing, this study chronicles a narrative of the end of empire that transcends the concerns of nationalism and resistance and instead focuses on the struggles to practice politics through the 1920s to the 1950s as the excesses and neglects of imperial governance gave way to the new politics of state-building and development. The transition to autonomous medical governance in the early 1920s created new varieties of colonial institutions and structures of government dictated by the demands of local politics. Rather than responding to the anti-colonial nationalism that shaped large-scale political projects, it was the local interpretation of issues of religion, ethnicity, language, class and caste that defined the parameters of health politics. This reading of interwar and early postcolonial governmentality therefore privileges local models of biopolitics, characterized by ambivalence towards anti-colonial politics and with a rigid adherence to the outcome of mainstream nationalism.

Until the twentieth century, public health policy did not officially engage with indigenous medicine. Instead, the Indian Medical Services (IMS) introduced a series of medical interventions that identified the Indian body, conceptualized as a vehicle for state-threatening disease, as a primary site of colonization. This changed in the twentieth century as the Government of India (GOI) began to consider the expansion of the public health agenda beyond the prevention and curtailment of epidemics. In so doing, it became evident that the GOI was very reliant on indigenous medical practitioners and began to engage directly with Ayurveda through the individual practitioners. By the mid-1910s, after several episodes in which it was clear that indigenous practitioners were crucial to securing the public health, the colonial government began to inquire into the training of said practitioners. Having dismissed Ayurveda previously on the grounds of its mystical provenance and unscientific logic, the government shifted its focus to the potential of the indigenous medical practitioner as a possible ally in the fight to improve upon the public health.

By the 1920s, Ayurveda was understood in both official and popular discourse as a set of practices mediated by vaids, or Ayurvedic practitioners, that reflected an ancient knowledge of the authentic Indian body but that was thoroughly applicable in the contemporary moment if administered by trained practitioners. This shift became more pronounced by the Montagu–Chelmsford reforms of 1919, which had introduced a new structure of dyarchic rule to the subcontinent. These reforms increased electorates and government positions for Indians and also transferred responsibility for medical services from the central government to the provincial government. In the United Provinces, these two developments came together to produce new problems for the growing state bureaucracy: the new lower-class employees of the provincial government had the right to claim sick leave, but they could not afford the fee that biomedical doctors charged to grant a certificate of ill health. The provincial medical board decided that these employees could get certificates from the practitioner they usually consulted, which initiated a debate as to the precise nature of indigenous medical practice in the provinces.

These discussions amongst elected and appointed provincial officials about the state of indigenous health also came to represent larger concerns about the state of the nation. After the Congress won the UP elections of 1937 – the first provincial elections in which political parties could run – the level of local state investment in the indigenous medical systems, and in Ayurveda in particular, rose greatly. Congress surveyed the extent to which Ayurvedic institutions were able to meet the population's medical needs, therefore instigating an official discussion about urban/rural/tribal populations and the responsibilities of government towards them. However, the politicization of Ayurveda also resulted in its involvement with North India's communal politics of the day. Muslim politicians felt that Unani medicine (associated, somewhat simplistically, with the Muslim community of South Asia, and cast as a minority tradition) was receiving short shrift in the new schema of proto-independent public health planning. Ayurveda also came to be associated in official political and lay popular discourse with decidedly Hindu articulations of political nationalism. In the years following independence, Ayurveda emerged as a universally 'national' medical tradition, properly modernized and fit to be governed, while Unani was marginalized from state planning and cast off as the cultural practice of a minority community.

This overview of Ayurveda in modernity reveals its role at the centre of several major shifts in governance, social structures and, ultimately, the postcolonial context for health planning. Equally important to the overview of Ayurveda's political history is the shifting perception of indigeneity and the body in popular culture. Ayurveda became an object of public interest through the myriad books, pamphlets and other texts, which detailed its content and practice. As such, it was intimately linked (along with other 'historic' traditions) to the development and expansion of Hindi-language literacy and the emergence of new modes of economic consumption, cultural production and social interaction. The nature of Ayurvedic discourse moved rapidly beyond the contours of the tradition as defined by its ancient texts, and instead was taken up with contemporary concerns about modernity. As such, the conversation incorporated 'modern' concerns of population, race and new medical techniques and technologies, thus bringing notions of progress and modernity into the realm of 'Vedic' tradition.

The earliest days of postcolonial planning began with a reflection upon the new possibilities for medical culture, both politically and socially, as an independent state government negotiated its responsibilities to its citizens. Fully adapted to the parameters of techno-modernity, Ayurveda was re-evaluated by its proponents who aimed to establish a cohesive and conclusive standard for expertise within it. At the state level, long lists were composed of those practitioners who could practice the tradition, who could frame progressive policy around it and who could instruct its teachings to future generations of doctors and scientists. In addition, new genres of Ayurvedic writing emerged, allowing little overlap or intertextual debate between the scientists, housewives, herbalists and doctors who each claimed to be authoritatively familiar with a facet – though not the entirety – of the Ayurvedic system of medicine. Gone was the older notion of a holistic set of practices rooted in ancient knowledge. In its stead, a system of interlinked but differentiated practices emerged, rationalized by the demands made of it by postcolonial regimes of health in a newly independent India.

Ultimately, Ayurveda was made modern as it was transformed to respond to the demands made by new experiments in governance – in essence, by its transformation into a governable system. This study takes its shape from this impulse: in it, I engage a variety of conversations about Ayurveda in the public and official domain to determine the scope of discourse, planning and, ultimately, the role of health in the broadest questions of governance. Other scholars have accounted for the internal lives of Indigenous Systems of Medicine through careful studies of practitioners, internal systemic change, medical discovery and the pragmatics of consumption.[4] This study builds upon these accounts of Ayurveda, Unani and local traditions and attempts to link them to broader logics of centralized Indian governance, first at the national and ultimately at the provincial level. The broader frameworks of health, community and the affective experience of tradition are interrogated and reconsidered in relation to the structures of power that deployed them in the logic of interwar and postcolonial governance. At the same time, a view from the periphery – be it from the perspective of the region, the minority community or a 'marginal' (and disciplined) population – reveals the fissures in normative, centralized structures of governance that can open up the spaces of critique. Taken together, the study will move beyond the grand-scale institutions of the centre and will take up the question of localized planning, in which the institutions built upon the missives of devolved power, and argue that it was precisely these institutions that sustained postcolonial development well beyond the transitory moment of independence. Rather than marking these institutions in relation to those of high colonial rule, this study will frame these innovations in governance and planning as always already at a remove from the more conventional logics of colonial rule, and steeped in the political possibilities of Indian governance outside of imperial determination.

## Biopolitical terrains: Medical history and the question of power

The challenge for the historian of medicine interested in non-biomedical systems in the subcontinent is the adoption of a historiographical framework, since Ayurveda has only very recently become a subject of historical investigation by medical historians. While Sanskritists and other religious scholars have been interested in the textual sources on medicine written throughout the ages, and anthropologists have addressed Ayurveda in the context of discussions of embodiment, consumption and ritual, historians have just begun to take up the project of historicizing Ayurveda. This reflects upon both the state of the field of contemporary medical history writing and also upon the tenacity of historical processes of marginalization that govern the way in which historians approach their subject matter.

Pre-colonial textual evidence about Ayurveda is a complicated reflection of the practice of medicine in daily life. Sources that exist from the pre-colonial period, outside of the context of Mughal or other governmental records, were often written in languages used to denote 'sacred' knowledge, and were therefore cut off from daily lived experience. Disciplinary divisions in the humanities often work to designate these sorts of investigations as fields of interest to religious or archaeological studies, creating an unnecessary division between fields. However, several new projects, some of which deal with the body, have taken steps to historicize these processes while relying on the skills employed in other disciplines to unravel their multiplicity of meaning.[5] Medical historians interrogating Ayurveda's post-classical moments have been particularly successful at mapping out textual developments within the historical processes of their contemporary moments.[6]

Historians of modern India working in vernacular languages and English face different challenges. The linguistic component does prove a barrier: while Sanskritists are quickly becoming historians, modern historians have not, for the most part, engaged with the study of Sanskrit or other 'dead' languages, as its practice fails to address the intellectual frameworks of colonialism and nationalism that have dominated the field of modern South Asian history writing. At the same time, the continued marginalization of Ayurveda from the institutionalized systems of biomedicine and public health that were used to govern South Asia, which we shall unravel in the course of the next two chapters, resulted in its virtual absence from the colonial archive. Colonial medical historians are generally careful to comment on its felt presence in medical encounters, but have not yet engaged with Ayurveda on its own terms.

Imperial medical historians have, however, used the presence of non-biomedical (though not significantly or purposefully biomoral) systems to gauge the reach of the medical arm of the imperial state. For example, Radhika Ramasubban has argued forcefully that the reach of the colonial public health agenda was limited and it left indigenous medical systems largely unaffected in the colonial period.[7] Public health, she argues, was closely tied to the army, to the imperial administration and to the associated white enclaves of the colonial state. However, others have problematized Ramasubban's approach. Arnold has argued against the 'enclavist' segregation of medicine, citing the aggressive push made by the sanitation commission for reform of Indian daily life from the 1860s, and the interventions made into areas generally not engaged with the state during the plague epidemic of the 1890s.[8] Harrison contests Ramasubban's claims by arguing for the contribution of indigenous peoples at the municipal level to decision-making about public health.[9] Ishita Pande's work, for instance, has provided a crucial interrogation of the roots of (il)liberal logics within the practices and policies of colonial medicine through a postcolonial reading of medical innovation in nineteenth-century Calcutta.[10]

Arnold and Harrison have implicitly called for an investigation of the state of the indigenous medical systems and function in the colonial period, a challenge which has been taken up in recent years by a new cadre of junior scholars who have taken up the topic of indigenous medicine in parts of the subcontinent which has resulted in the proliferation of a set of themes and conversations about indigenous medicine. These studies illuminate the ways in which Ayurveda and Unani medicine were transformed internally to respond to some of the demands made by the new market in medical products and the new norms around knowledge production introduced by colonial medicine. Kavita Sivaramakrishnan's *Old Potions, New Bottles: Recasting Indigenous Medicine in Colonial Punjab*, the first of the recent studies to emerge in published form, evaluates the division of indigenous medicine into specific categories (Ayurvedic, Unani and Punjabi Baidak) tied to community, caste and religion, and traces the evolution of Ayurveda in

particular throughout the nineteenth century in response to the introduction of colonial biomedicine.[11] Guy Attewell's *Reconfiguring Unani
Tibb* takes a more conceptually rigorous approach to the study of
indigenous medicine, positing a series of questions about the systematization of indigenous medicine, and tracing a reconfiguration of Unani
in the nineteenth and early twentieth century that parted ways with
the conventions of the past and carved out a medical modernity for
Unani that took up the particular concerns of community and religion,
and that was characterized by the heterogeneity of the set of practices
that came to compose the tradition.[12] Seema Alavi's *Islam and Healing: Loss and Recovery of an Indo-Muslim Medical Tradition, 1600–1900*
uses the long history of Unani Tibb to gauge the cultural transformations of Islamic life and culture throughout the colonial period, but
framing its development against the backdrop of the declining Mughal
court patronage and attempts by hakims to model their reform in the
nineteenth century based on community-based forms of knowledge
production and professional organization, rather than strictly in relation to Western biomedicine. Finally, Projit Mukharji's *Nationalizing the
Body: The Medical Market, Print and Daktari Medicine* reorients the lens
of historical investigation to nineteenth-century Bengal and traces the
Bengali response to Western medicine, and frames a 'daktari' tradition
composed by the experiences and experiments of Bengali doctors who
responded to, adopted and ultimately translated Western medicine for a
Bengali audience.

Taken together, these texts identify a set of themes and practices
that are beginning to define a methodological approach to the history
of indigenous medicine in the subcontinent that will have resonance
in other colonial contexts. The focus on region, on vernacular language and on the particularities of community, caste and religion has
worked to decenter the colonial state, therein complicating the model
invoked in the works of Ramasubban and other imperial medical historians. Where imperial historians have begun by positioning indigenous
medicine in relation to biomedicine, historians of indigenous medicine
have instead cast imperial medicine as a factor (albeit a large one) in
line with many other concerns, trends and movements that influenced
the way in which these traditions were transformed and reconfigured
in modernity. Alavi's work, for instance, argues that the reinvigoration
and recomposition of Unani Tibb is the product of the radical transformation of Islamic knowledge production in the nineteenth century –
itself the product of globalizing Islamic cultural and intellectual identity
throughout many imperial and extra-imperial spaces. Mukharji's work,

which does take up the question of colonial medicine directly, swiftly moves beyond a simplistic dichotomy of Western-versus-indigenous and instead reveals the rich workings of a Bengali society that is itself becoming modern. Ultimately, the transformation of different forms of indigenous medicine and body knowledge into systems that could be deployed and relied upon to tackle modern ills are well characterized as a complicated mode of mimesis, in which the techniques and technologies of allopathic medicine – along with the logic of knowledge production about the body, the commodification of medicine and the professionalization of practitioners – provided guidelines for the transformation of local health contexts.

A gap remains, however, between the history of indigenous medicine as constructed in these particular contexts and the greater rubric of health, empire and politics that has been the concern of imperial medical historians for some time. The studies of indigenous medicine provide a wonderful starting point from which to engage with the broader history of health in the subcontinent, especially in the later years of empire; few, however, take up the question of the colonial state beyond the immediate concerns of the local context in which the study is framed.[13] This leaves a lacuna in conceptualizing the broader relevance of these studies to the histories of both empire and its postcolonial decolonization. The impetus for my study is thus the bridging of this gap: between region and centre, between community and nation, and between medicine and health. In so doing, I aim to bring the histories of indigenous medicine back into line with other very powerful attempts to interrogate the history of health in the subcontinent, and to bring the history of medicine – even, or perhaps especially, in its most diverse and traditional iterations – to the centre of a political history of India. Following in the footsteps of other historians of health and the body, this intervention begins by reframing the relationship between subjects and purveyors of medical experimentation through an analysis of power.

The question of power has preoccupied historians of health and the body for some time. Imperial historians often harken back to a vexing question posed by the African medical historian Shula Marks at an early conference on medicine and empire: 'what is colonial about colonial medicine?' In so doing, Marks challenged medical historians to think about the unifying factors that connected one colonized region of the world to another as the specific, regional histories of medical life were presented.[14] Marks attempted an introduction of an inherently transnational reading of the history of medicine, which would recast it within its wider networks spanning the globe that colonial medical interventions and inventions helped to produce. My concern is precisely the opposite, constituted as it is within the realm of medical systems labouring under colonial hegemony but also extant in other spaces. However, I still find Marks' intervention a crucial one for my purposes, as she provides an intellectual framework through which historians can consider the medical experiences that occurred under colonialism but occupied an ambivalent position vis-à-vis the colonial state.

Building upon this theme, Warwick Anderson has questioned the continued use of colonialism as a mere historical marker rather than a relationship of power.[15] This introduces the second major trope at work in an investigation of medical practices extant outside of the scope of imperial medicine: the evaluation of colonial medical institutions and practices as both representations of and the means through which power was negotiated in the subcontinent. David Arnold's groundbreaking work *Colonizing the Body* remains the most thorough exploration of colonial medical practices in India shaped through this lens.[16] In it, he adapts a Foucauldian approach to the history of the body – and its relationship to institutions and practices of state control – to evaluate the practices of the state vis-à-vis the colonized subject, exploring in particular the ways in which 'colonialism used – or attempted to use – the body as a site for the construction of its own authority, legitimacy, and control'.[17] Most medical historians writing after Arnold have adopted this approach, especially with regard to the emergence of the critical study of race as a category of historical analysis. As Sarah Hodges argues, ' "race" was not only constituted in large part by colonial power, race was a key sign through which colonial power operated'.[18] Studies of the pathologizing and disciplining of colonial sexualities have also relied upon the questions raised by Arnold, but have succeeded in pushing the boundaries of his study by adding a gender dimension to an exploration of power.[19] Taken collectively, the repositioning of the body and the intimate practices that sustain it at the centre of the colonial project provides a useful way to think through the refashioning of Ayurveda to suit the needs of the self-fashioned nationalist subject, which we shall explore further in Chapter 3.

Pervading these conceptual modes is the relationship between health and the nation in the modern period. This is a crucial connection outlined by Gyan Prakash in *Another Reason: Science and the Imagination of Modern India*, in which he moves beyond the foundation of power through the institutionalization of science as a colonial practice and instead focuses on the role of 'science's cultural authority as the legitimating sign of progress and rationality' in the age of nationalism. Taking this as a basis, Prakash traces the reworking of genealogies of 'traditional' systems of meaning through the lens of scientific language, resulting in the emergence of themes like 'Hindu Science', which worked to serve the nationalist project. This approach resonates with attempts to historicize medical traditions lauded for their ancientness in terms that carry the weight of legitimacy in the contemporary moment, which, as we will see in Chapter 3, was always the struggle for those wanting to make Ayurveda relevant to the modern world.

More recently, Sarah Hodges and Stephen Legg have both expanded upon this intervention by further fleshing out the biopolitical imperative in Indian history in the interwar context. Where Arnold identified the role of the state in the configuration of power between 1857 and 1914, Legg and Hodges have pushed the paradigm by undertaking more nuanced readings of the notion of power in its diffuse and varied forms. In a study that takes the birth control campaigns of South Indian reformers in the interwar period as its point of inception, Hodges maps out a late colonial biopolitics hinged upon the articulation of the 'population problem' in colonial Madras.[20] Hodges' pivotal intervention is in exploring the notion of the political outside of the realm of the state, a case she makes through discursive engagement with the campaigns to promote birth control and to curb undesirable populations led by laypeople organized into clubs, societies and non-state organizations during the interwar period. Steven Legg, returning to the realm of the state, refocuses the Foucauldian paradigm to broaden the spectrum of biopolitics by exploring the larger effects of governmentality, through the convergence of sovereign power, disciplinary biopower and governmental power once again around the question of population. Legg's work on the role of population in the planning and management of the combined space of New and Old Delhi reifies the idea of power as a multi-sited affair.[21]

These approaches elevate health governance to newly sophisticated realms and make evident what is most useful about biopolitics as a conceptual framework for the history of India. Fundamentally, a late colonial biopolitics locates power at a confluence of interlocking concerns and administrative operations, initiated and carried out by a variety of actors, centred on the management of the body and the problems and possibilities it represents. The interwar period also represents an epoch in which the colonial state began to withdraw its intimate involvement in the political life of the subcontinent, acting as a satellite force in competition with localized governmentalizing forces on the ground. The 'problem' that the indigenous body posed to state and society was conceptualized differently by a wide variety of actors situated along various political axes but converged around questions of resources, space and the possibilities of postcolonial governance.

There is also a need to use the interwar experience of health governance to push the biopolitical schema into the creative and hybrid spaces of late colonialism. Colonial historians of sexuality and the body have denoted the paucity of a more sophisticated discourse of subjectivity and subjecthood in models of the biopolitical, most notably vis-à-vis the categorization of race and racialization.[22] The Indian interwar period submits the original model to further pressure as it requires an exploration of subjecthood that takes relationships of power that aggregate around systems of racialization.[23] The late colonial state poses a quandary for historians of race and the body: when viewed from the margins, whiteness remains an important and foundational marker of power, but the daily experience of subjectivity is much more urgently and profoundly shaped by caste and religion. While elites are likely more successfully marked by their mobility and access to both imperial and cosmopolitan whiteness, the actors within the local structures of governance – especially its lower rungs, where the governance of Ayurveda was enacted – were confined by what tends to be understood as 'racial ordering'. Instead, the bodies in question – both in terms of enacting and being disciplined by governance – were much more meaningfully marked by indigenous structures of embodiment that were in dialogue with white supremacy but that envisioned new loci of power and authority along lines of class, caste, religion, ethnicity and region. At the same time, the affective invocation of 'tradition' worked to further inflect Ayurvedic legitimacy with the trappings of Hindu symbology – figures of the pandit, the scroll, the indigenous authentic. Taken together, the discourse of population management, of disciplined bodies, of the rule of the biometric and the technocratic is necessarily hinged on and evocative of questions of power and subjectivity. It is not any body that is acted upon – it is the Muslim, the *shudra*, the *adivasi*, the *Dalit*, the rural body that figures prominently that marks the foundational building blocks of discourse.

The interwar period in India provides a compelling working ground through which to envision a multiplicity of biopowers because the ontology of power was itself so variegated. The Raj, the clear force behind the biopolitical structures of the nineteenth century, was in a period of retreat. The everyday arm of the state became less infused with its former reach, becoming a satellite presence of surveillance, while also leaving ontological gaps that could be filled by a variety of new actors.

The local sanitation societies of Hodges' study, the authoritarian vaids and other health writers we will meet in Chapter 3, the provincial councils developed under post-1919 dyarchic structures of governance and, finally, the ruling committees of Indian National Congress of 1937–1938 were all apparatuses of power in this period. What underlies these actors and their representative apparatuses is a consistent framing of power through duelling notions of population, thus creating, following the Foucauldian missive, a connection between the heterogeneous elements that constituted the category of urgent need.[24]

While a moment of creativity in planning proliferated from the early 1920s until 1935, the advent of a newly formalized structure following the Congress Party's electoral gains in 1937–1938 recast the question of population as a central focus for health governance. As we shall see in Chapters 4–6, the surveillance and management of the populations of the United Province (later, Uttar Pradesh) was in part conceptualized through the ways in which Ayurveda was modernized. More plainly writ, Ayurveda's ability to adopt biopolitical practices formerly associated with colonial medicine became a benchmark of its march towards modernization, as well as the blueprints for a teleological plan towards its continued improvement.

## Conflicting modernities: Culture, society and politics in interwar North India

Studies of the biopolitical reveal the specificities of Ayurvedic politics and practices, casting an ahistorical framing of Ayurveda as naïve and simplistic. However, the affect around Ayurveda still invokes, at the very least, questions of 'tradition', especially via the idea of encounter with Western biomedicine and scientific rationalism. The extant narrative of colonial encounter itself relies heavily on the drama of the clash, and even away from the extremities of the neo-cons, Indian historians have long framed encounter through debates of either continuity or disruption.[25] The historiographical drama of the clash is perhaps felt most strongly in the productive tension between the different schools involved in the debate, which resulted, for instance, in the post-structuralist and postcolonial turn towards the subaltern and new considerations of the operation of power in the colonial world.[26] Regardless of the 'school', the overall tract remains reliant on an almost chronological conclusion of outcome: predicated on the 'moment' (or moments) of encounter, it continues to situate the journey into modernity as a historically earlier one, staked on the disruptions and continuities of eighteenth- and nineteenth-century contexts of political economy and cultural life. As such, the poles of Orientalist cultural production, the question of Asiatic despotism, the imposition of liberal institutions and the introduction of new regimes of embodied discipline (among many, many others) meet a set of critiques aimed at qualifying resistance, categorizing colonial violence, locating indigenous agency and participation and identifying the multi-sited capillaries of Imperial power.

How should Ayurvedic encounters with modernity be conceptualized after 'encounter' has become, if not resolved, then both commonplace and banal? As we shall see, Ayurveda overcomes the gloomy spectre of colonial legislation against it in the 1910s and becomes a cornerstone of state health planning after the First World War. Rather than coming into a colonial modernity concerned mainly with Orientalist constructions of Indian tradition, Ayurveda instead comes of age as a trusted salve for the political demands made of health traditions in a new era of diffuse, devolved and localized politics. The affect of clash is replaced by the pragmatics of utility.

The model for evaluating the politics of interwar Ayurveda must begin with a reading of the particularities of its context. This study takes the United Provinces of Awadh and Agra – later, Uttar Pradesh – as its focus, though the themes explored have greater resonance for some of the princely states and other provincial territories of North India. The United Provinces provide a good backdrop for many histories central to the unfolding of social processes. Most pragmatically, political activity in the UP was at the fore of nationalist agitations during the last decades of colonial rule, and the ascendancy of the Congress government in this geographical and political context serves as an important model for understanding the importance of communalism and nationalism in South Asia.[27] From a medical perspective, UP was also a site of great innovation because its state Medical Board pushed through legislation on the indigenous medical systems that set the model for other provinces, as we shall see in Chapter 5. Taken together, these perspectives on the state and the communities within it will allow a reading of Ayurvedic tradition, signified as the ancient Hindu 'truth' about the body, through the variegated politics of the contested terrains of the United Provinces during moments of transformation and transition.

The interwar period requires a new set of parameters for describing the resurrection of Indian tradition, one that pays heed to the well-crafted models of 'tradition and modernity' set in high colonial times but that foregrounds the historical moment of the interwar modern. The dyarchic division of power of 1919 led to a refraction of colonial power, characterized by the aforementioned 'excesses and neglects' of the colonial state, but much more powerfully by the growing importance of semi-autonomous provincial power.[28] Traditionally, scholars of Indian history have decried dyarchy to be a massive failure (we'll revisit this idea in Chapter 4) and instead have used this moment of political cohesion to think through the impact of new ruling elites on the structure of Indian society.[29] More recently, scholars working on interwar structures of government have countered this reading by exploring the illiberal machinations of the state as it overstepped its bounds in the provincial realm (namely, through moments of disciplinary violence against anti-colonial protest).[30] Scholars working through an internationalist perspective have decentred a grand narrative of colonial repression/anti-colonial protest by charting the rich engagement with trans-imperial, trans-national and international movements and phenomena.[31] Other terrains of historical engagement with the politics of the time will, in the future, no doubt trace the role of independent actors, working both through the state and outside of it, who looked to profit from the loosening of control over Indian industry and agriculture; these studies might range from early developmentalist interventions from international agencies (perhaps traced neatly through genealogies of both aid and neoliberalism in the subcontinent) to that of individual entrepreneurs or even to those working with and through the League of Nations (of which India was an independent member).[32] Moving back to the realm of the local, histories of sexuality have identified both the flourishing of localized, vernacular discussions of the body and its functions through actors and groups very much engaged with the broader scene of sexology, eugenics and global population concerns.[33]

However, the United Provinces was also the site of some of the major political contestations on the nature of community and of the nation. This is true of both the political and cultural sphere. Historians like Gyan Pandey, Francis Robinson, Paul Brass and Lance Brennan have explored the rise of communal politics through the experience of Muslims in the nationalist politics of the UP.[34] Despite their different approaches and arguments, they all agree upon the fuelling of a division between Muslims and Hindus by the colonial government, by cultural revivalists, religious extremists, by educational and cultural movements and institutions, and by those looking to rearticulate collective movements formally drawn on ethnic lines. Literary, cultural and social historians have also identified UP as a central site where the boundaries of the nation were drawn. This was due in part to the concretization of identities through linguistic and broader demographic change that led to social upheaval, the decline of the former authority of Islamic culture in Awadh and the emergence of new groups into the public sphere. William Gould's work has linked these two spheres together, by exploring the importance of popular discourse to the realm of high politics, which blurred the lines between the popular and official mode of discussing the nation, especially through the rhetoric of the Congress Party.[35] The importance of print culture to the self-fashioning of individuals as modern beings and to the cementation of new ideas about politics, culture and identity has been identified by historians and literary critics as a key trope in the study of India during the late colonial period. Literary scholars have turned towards the vernacular print cultures of the late nineteenth and twentieth centuries to trace social and cultural histories of the formation of nationalist identities in this period.[36] Historians have also relied on the field of vernacular literature to gauge the self-fashioning of authority by new groups who were financially solvent and socially powerful, which Sanjay Joshi has referred to as the 'project of fashioning a middle class'.[37]

Any investigation of the connection between culture and politics in this period begins with Vasudha Dalmia's seminal text on Bharatendu Harischandra, one of the most eminent Hindi literary figures of the modern period.[38] In it, Dalmia traces the emergence of a standard, uniform language known as modern Hindi, written in the Devanagari script, and links the formation of language to the creation of national identity. Thus, Dalmia argues that language was vested with the responsibility of representing the nation because the progress of language came to represent the progress of the nation. Francesca Orsini begins where Dalmia ends by looking at the reform and consolidation of the literary sphere and publishing world into what she calls the Hindi public sphere in the 1920s–1940s.[39] Orsini shows how the growth of the public sphere and the incorporation of other groups such as women and lower-caste groups did not lead to an increase in the divergence of opinions but instead helped to spread normative ideas about the nation to new groups. At the same time, the expansion of the public sphere to include women's writing and the writings of other caste groups reflects the expansion of middle-class dominance more than it symbolizes a growing diversity within the participants. Orsini also stresses the role of language being a vehicle for nationalism, as it is further standardized, 'purified' and applied to the nation beyond its heartland in UP. The Hindi public sphere ostensibly became the sphere in which new ideas about the nation were debated and discussed, and in which modern national identities were negotiated. Alok Rai, in a study of the intersection of nationalism and language, further underlines the connection between identity formation and literary expression by looking at the potentially – and sometimes rather overtly – communalist underpinnings of these developments.[40]

Charu Gupta further complicates this take on the public sphere by exploring the negotiations of taboo subject matter in the Hindi public sphere.[41] Using the rubric of 'obscenity', she argues that literature was employed to patrol the boundaries of the nation, especially through the models of appropriate gender performance and sexual identity reified in the public sphere. Gupta argues that the Muslim man was vilified while the Hindu woman was held up as the ideal of appropriate nationalist identity. Moreover, the shaping of a class identity around these poles of appropriate social interaction makes evident the more social uses of the Hindi public sphere.[42] The rise in power of new social groups who formed alliances in the new spaces of modernity, like educational systems or social clubs, resulted in a formation of relationships outside of the context of kinship or *jati*, and also resulted in a reification of the ideals of modernity as a basis for forming community. Part of the performance of this new middle-class identity was participation in the public sphere through intellectual debates on the social issues of the time. The middle-class perspective therefore dominated the public sphere because the 'project' of making the middle class was heavily reliant on the evolution of the public sphere. Hence, the middle class, as a social group, was most invested in it.

The dominant idea that emerges from this collective writing on the public sphere in its social context is the reification of the Hindu middle class as a coherent category. It is tempting to follow a Habermasian interpretation of the public sphere as a space where formally marginalized groups can comment on social norms and policies of the state and thus broaden the scope of debate. In fact, scholars have found quite the opposite – that the public sphere contributes to a normalizing and consolidating of the diversity of meanings relating to any specific concept, including what it means to be a modern Indian.[43] It is through the middle-class domination of the public sphere that certain interpretations of a concept like Ayurveda become privileged over others, and thereby become the uncontested norm.

Taken together, these studies interrogate the 'natural' formation of Indian nationalism, rendering complex the easy trajectory that puts liberal nationalism in line with anti-colonial resistance as the unchecked moral framework of the modern Indian ethos.[44] Of note is the care

with which these scholars, working across disciplines and methodologies, depict a complex multiplicity of nationalisms that undo the
representation of a collectivized, universal nation positioned against
the empire, and engaged in a singular, anti-colonial struggle. There
are, of course, moments of mass collective struggle that need to be
remembered. However, historians of the twentieth century have broadly
revamped 'nationhood' as an ideal that lent strong idiomatic appeal to
the particularities of community; rather than one nation under Gandhi,
late colonial/high nationalist India is much more productively characterized by the rigidification of identity along community – and, often,
communal – lines. A more universalist reading of these national exceptions might propose a process of identity formation whether with or
against the national. Instead, recent work by historians working on
specific communities reveals careful plans to carve out collective identities on their own terms. Research into Muslim cultural production in
the UP, most notably by Justin Jones and Markus Daechsel, reveals a
rich cultural life engaged with modernity and the modern that complicates the dominant 'national' model suggested in studies of Hindi
and Hindu community, and those proposed by shallow readings of
Islamic counter-nationalisms. By similar measures, Sarah Beth and Badri
Narayan have illuminated *Dalit* worlds of politics and cultures that take
issue with these dominant readings as well, referencing the icons and
images of mainstream Hindu nationalism, while suggesting localized
readings of the *Dalit* historical that radically reinvent the trajectory of
the Indian past.[45]

This analysis of the distortion of modern regimes of power provides an
excellent model through which to evaluate the processes of grand-scale
state building. To return to the question of modernity after the clash of
encounter, the gift of the interwar political schema is the possibility of
understanding the provincial political as more than just a refraction
of the high national. This is particularly crucial for the period of civil
unrest and violent extremes that have come to define our discourses
of communal politics, and rightly so. The partition of the subcontinent represents the largest migration of people in human history, and a
legacy of violence that has shaped the outcome of subcontinental, and,
indeed, global politics well into the twenty-first century. To minimize
it is to erase the foundational moment of South Asian postcoloniality, not to mention the culmination of centuries of colonial violence
and the most macabre extremes of resistance to it. Mythologies of high
nationalism and partition frame the late colonial state as incapable of
functioning, felled by the success of wartime protest against the empire

and by the dysfunction and chaos of violence and migration. Indeed, the postcolonial reader measures *hartals*, *bandhs*, boycotts and hunger strikes by the ability of resistors to shut down imperial business as usual.

And yet there is a banality about the everyday that must also be acknowledged, in which communalism was equally implicated in a less overt fashion. This is where the experience of interwar and late colonial local politicking can further complicate our approach to histories of communalism: while the violence at the border and in the major urban centres disrupted everyday life, hospitals and dispensaries in local districts continued to be built, members of the Board of Indian Medicine continued to show up for work and to survey the institutions under their purview, classes in Ayurveda and Unani medicine continued to meet, contracted officials continued to survey and commissions continued to plan. The state, in its nascent, proto-postcolonial and localized form, continued to run. How then do we understand the continuity of daily life at moments of extreme politics in a way that does justice to the historical legacy of violence?

This study, by way of redress, examines the practice of everyday politics to reveal the genealogies of communalism that lay at their roots. Ayurveda as cultural practice remains firmly within the realm of the Hindu, the casted and the classed. It is culturally mainstream (despite the alleged rarity of practice) and easily emblematic of the glory days of Hindu civilization. It is shrouded in the nostalgia felt for that imagined epoch, while its thorough absorption into the structure of modern governance allows the necessary distance that produces this variety of nostalgia. It is for this reason that the historical and political contextualization of Ayurveda is crucial. While Ayurveda's political history – and most of the actors involved therein – is far removed from the extreme moments of communal violence, the continued work undertaken by local governments to keep Ayurveda Hindu and to maintain the Hindu traditional as the only possible cultural norm links this study intimately with others that foreground moments of communal conflict.

Finally, this study re-examines a conception of power and authority held fast by historians interrogating the conditions of postcolonial state-building and wanting to cast aspersions as to the nature of independent India's political modernity. Partha Chatterjee's now familiar argument about the nation and its fragments begins with the claim that it is not analytically productive to distinguish forms of the modern state from the colonial state. Instead, Chatterjee argues for an understanding of the colonial state that situates it at the heart of the narrative of political modernity through the apparatus of power it introduced, and which are still at play.[46] Chatterjee argues that the modern state in India has merely rearranged and expanded the institutional framework of the Raj, rather than transforming entities like the law, the police, the army and the bureaucracy. I am both sympathetic to and aligned with the political impetus that Chatterjee goes on to conceptualize as the rule of colonial difference, especially for the critique of a liberal universality that was continually promised and ultimately deferred by the Raj. Liberal British administrators promised that colonialism would eventually eradicate difference by bringing colonized people forward into the fold of progress and history; once colonized 'others' became modern subjects, colonial control would have lost its ideological foothold. Instead, the outcome of Indian history unfolded is the legacy of institutions entrenched in the flippancies of liberal rhetoric, and subjected more pragmatically to the pulls of identity and corruption, rendering the postcolonial subject to the flaws and inequities that characterized colonial liberal governance.

As powerful a mode of analysis as this remains, it has the potential to become (if only in shorthand) a reading of a hegemonic colonial state characterized by symmetrically uniform institutions – a reading that loses the nuance and diversity of the specifics of institutional development and leaves little room for the refutations, resistances, ambivalences and collaborations of state institutions with the larger rubric of colonialism. At the same time, there is a parallel story about institutional development that might help to reinvent this argument for a new realm of governance. The latter chapters will explore a range of state-sponsored and provincially administered institutions that come up in the context of dyarchic governance, when the reach of the Raj is at a remove and the everyday responsibilities of government to its proposed citizens alters the experience of colonial governmentality. They will also explore the ways in which the institutions that sustain an easy administrative transition to independence are precisely these boards, councils and committees that come about once the provinces of British India become more responsible for the everyday lives of their citizens.

*   *   *   *

The first two chapters of the book locate Ayurveda in modernity. Chapter 1 grapples with the challenges of historicizing Ayurveda through an exploration of the conceptual approaches needed to construct a genealogy of its development. Ayurveda is best approached through a variety of methodological lenses, and this chapter will consider it in light of four intellectual perspectives: Indological, anthropological, literary and historical. It goes on to discuss Ayurveda as a biomoral tradition, informed by anthropological readings of the notion of biomoral exchange. Finally, it reflects upon the early colonial history of the indigenous medical systems so as to explore the reflexive relationship between Ayurveda and colonial modernity at the turn of the twentieth century. Chapter 2 explores the moment of transition from the colonial state's exclusion of Ayurveda, from planning to the gradual attempts to address the legitimacy of the system in modern terms through. These exploratory gestures made by the Medical Department revealed the extent to which indigenous practitioners were often relied upon locally to implement a centralized public health agenda and that myriad networks, relationships and experiences of public health policy were heavily reliant upon indigenous cultures of physicality, practices of extra-biomedical medicine and pre-colonial patterns of medical consumption.

Chapters 3–5 explore the transformation of Ayurveda into the public and political spheres, charting its evolution in discourse and policy as it came under the rubric of governance. Chapter 3 explores popular understandings of Ayurveda, focusing on Hindi-language discussions of new medical techniques, technologies and conditions that had been previously left unexplored within its purview. This chapter explores specific genres of writing and evaluates the ways in which Ayurveda came to be understood in popular culture. Serving as a 'scientific' explanation for the constitution of the Indian nation, Ayurveda emerged as a decidedly amorphous signifier, divorced from historical trajectory or bounded meaning, and was employed as a broad category signalling an indigenous approach to questions of belonging and difference. Chapter 4 shifts to the sphere of formal politics by charting the incorporation of the Indigenous Systems of Medicine into the revamped provincial medical services following the 1919 transition to a dyarchic structure of governance, through a set of reforms that extended the franchise and introduced portfolio-holding positions for elected ministers at the provincial level. Chapter 5 traces the process of institution-building that characterized the expansion of the medical infrastructure following the Indian National Congress' electoral victory of 1937 to the early decades of independent planning in the United Provinces. Ultimately, these latter chapters chart the ways in which Ayurveda was regularized as a suitably modern system suited to model of health through a biopolitical engagement with population management and the first inklings of a developmentalist ethos in India.

Finally, Chapter 6 considers Ayurveda's inclusion in an imagining of a new modernity for India, where the missives of progress and economy were reconfigured along the fissures created by various stakeholders in the independent Indian nation. As such, this chapter argues that, at the beginning of what was to become an era of experimental planning on a local, national and global scale, an intimate familiarity with Ayurveda itself became a designation of an attractively authentic and thoroughly modern understanding of the postcolonial Indian condition. Working through several genres of contrasted knowledge production about Ayurveda in the 1950s, this chapter reveals a plurality of stakeholders, ranging from federal politicians to state bureaucrats to educational instructors to lay authors, who scrambled to lay claim to an authoritative knowledge of Ayurveda, and who together formulated a politics of the possible for Ayurvedic medicine.

# 1
# Historicizing Ayurveda: Genealogies of the Biomoral

Over the course of ancient and post-classical Indian history, 'Ayurveda' evolved from a textual term for the knowledge of life into a medical tradition with a literary canon, recognized health practices, and practitioners asserting their expertise and expecting elevated social status. The pre-colonial development of Ayurveda reflected a holistic approach to the natural world, uniting beliefs about the physical structure of matter with metaphysical and religious insights. In Ayurvedic texts, medicine is closely associated with philosophy and ethics; similarly, medical practice was located within a wider context of ritual and social behaviour. Ayurveda gained coherence and influence as a collection of medical practices that were in harmony with, and indeed reinforced, both Sanskritic learning and the structure of the Indic societies that sustained it.

Consequently, the assertion that Ayurveda is an Indic and Sanskritic tradition is more than just a statement about its geographical and linguistic heritage. The processes through which Ayurvedic medicine was consolidated influenced later developments for two principal reasons. The first is that the creation of patterns for determining what constituted authentic Ayurveda resulted in the establishment of channels of textual, political and economic legitimation for practitioners and practices in the modern period. Ayurvedic practitioners came to stress their medicine's unbroken lineage from the Vedas and from ancient practice, emphasizing in particular its Aryan, Sanskritic and Hindu roots. A second reason for dwelling on the pre-modern history of Ayurveda is that, as a body of medicine, it actually drew on a multiplicity of sources and influences. This hybrid history not only reveals the inadequacy of Hindutvic accounts of Ayurveda but also points to the interactions and overlaps between several different medical traditions that came to necessitate a more aggressive demarcation of Ayurveda in the modern period.

This vague delineation of Ayurveda as caught somewhere between science and faith is encapsulated in the discourse of medical historiography through the identification of Ayurveda as a biomoral tradition. At its most simplistic, this label is employed to differentiate Ayurveda ideologically from allopathic or 'Western' biomedicine. The biomoral becomes a catchall phrase solely marking its distinction from Western medicine, with vague references to the moral frameworks of embodiment that characterize South Asian life. The applied association in this literature is that Ayurveda has something to do with religion, or the realm of the spiritual or transcendental, thus moving it firmly out of the scope of rational medicine. In essence, the possibility of Ayurvedic adherence to a logic that moves away from the evidence-based medicine forms the ground of the system's dismissal from taxonomies of rational science, a categorization that rendered it 'inappropriate' and 'unreliable' in the context of medical modernity. The biomoral parameters of Ayurveda can be deduced precisely, however, and oriented around several of the major moral questions associated with the social, cultural and political life of South Asians.

This chapter thus attempts a historicization of Ayurveda in both theory and praxis, dynamically located both within the historical record and also within the conceptual imaginings of social relations in the subcontinent. It begins with an overview of Ayurveda's evolution through realms of textuality in the Sanskritic tradition, highlighted to reveal its foundational linkages of embodied medicine to the broader evolution of Hindu political life in the subcontinent. It then moves towards the realm of the conceptual in a discussion of the evolving notion of the biomoral in anthropological thought, in which the vague category of 'religion' is replaced by the demands of social relations and, ultimately, the inevitability of biopolitics. Finally, it explores the political possibilities of the biomoral in action, framing Ayurveda's deployment in colonial times and arguing that Ayurveda came into modernity through a political articulation of embodied indigeneity. Taken together the genealogy of biomoral possibilities reveals the foundational political impetus that frames Ayurveda's rich and complex history.

## Ayurvedic systems in time and space

The antiquity of the Ayurvedic tradition in Sanskritic literature has been an important source of cultural and medical authority for practitioners.

It acquired particular importance during the Orientalist resurgence in Vedic scholaiship in the early nineteenth century. Ayurveda is derived from the terms *Ayus*, life, and *Vidya*, knowledge. Although there are references to medical science in the *Rig Veda*, which appeared circa 1200 BCE, the first literary codification of a medical tradition is thought to have come much later in the final chapters of the *Atharvaveda*, the fourth Veda, which is the oldest Sanskrit text to deal with the physical sciences at length. This text describes the use of poisons in warfare, gives a list of therapeutics for certain diseases and attributes certain diseases, including leprosy, to external infectious agents.[1] Ayurveda as described in the *Atharvaveda* and elsewhere is considered to be an *upaveda*, a secondary Veda, putting medical practice at the centre of Vedic thought.[2]

This ancient pedigree was much emphasized by medical practitioners who wanted to support their status in society with the weight of Vedic authority. For instance, Debiprasad Chattopadhyaya has used the *Rig Veda* to argue for the eminence of practitioners in the Vedic period.[3] However, the information presented in the *Atharvaveda* weakens practitioners' claims to Vedic authority, since the first text to deal with Ayurveda at any length actually minimizes the role of practitioners in the dissemination of medical knowledge.[4] Moreover, scholars have argued that there are important discontinuities between the fragments of medical knowledge contained within the *Atharvaveda* and the dominant medical traditions that evolved into classical Ayurveda.[5] This raises the question of how far the *Atharvaveda* can be considered a reliable guide even to contemporary practice. Dominik Wujastyk argues that the system of humours (*dosas*), which is a central principle of classical Ayurveda, is not mentioned in the *Atharvaveda*, nor are the other constituents of the body with which these *dosas* interact. However, the cultural historian A.L. Basham identifies in this textual history larger social and cultural shapings of medical traditions by contemporary values.[6] Regardless of the medical details, the cultural ethos of the tradition is inherently linked to the social structures and cultural mores that emerged in the classical period.

This tension between textuality and social mores characterizes attempts to historicize traditions of the ancient Indian past. A canonical literary tradition with strong ties to later Ayurvedic practice began with the *Caraka Samhita*. Though attributed to the scribe Caraka, it is most probably the effort of several scholars who worked together to compile its content. It was composed in the first decades of the second century CE, during the reign of Kanishka. The text is framed as a discussion between the sage Atreya and his pupil Agnivesa and is divided into eight sections: *sutra* (rules), covering pharmacology, food, diet, some diseases and treatments, physicians and quacks, and varied topics in philosophy; *nidana* (causes), describing the causes of eight main diseases; *vimana* (arrangements), discussing various topics such as taste, nourishment, general pathology and medical studies; *sarira* (relating to the body), treating philosophy, anatomy and embryology; *indriva* (the senses), describing diagnosis and prognosis; *cikitsa* (therapies); *kalpa* (pharmacy); and *siddhi* (completion), outlining further general therapy.[7] While the text has often been considered to embody an exclusively Indian tradition, the similarities with Chinese, Persian and Greek medicine are noteworthy.[8]

The second major canonical text of Ayurveda is the *Susruta Samhita*, which follows the model put forth by the collators of Caraka's text. Though a grammatical rule recorded in 250 BCE mentions it in passing, a re-edited version of the compendium is thought to have been produced in the sixth century CE.[9] It takes the same lyrical form as the *Caraka Samhita*, in this case recounting a discussion between Dhanvantari, a king of Banaras, and his student Susruta. Dhanvantari's name came from the term *Dhanuh*, or surgery, and a grammatical shaping of the name to imply that he is fully skilled in it.[10] Despite a lack of evidence about Dhanvantari beyond this source, his name has become associated with the practice of Ayurveda, and many contemporary texts pay homage to his perceived wisdom. The sections of this work are similar to the *Caraka Samhita*, omitting only the *vimana, indriva and siddhi*, but including a section called *Uttara* ('last'), covering ophthalmology, the care of children, diseases ascribed to demonic attack, dentistry and aspects of medicine not dealt with elsewhere.[11]

The enduring centrality within Ayurveda of the principles outlined in the classical texts ensured that mastery of their content, familiarity with Sanskrit and appeals to ancient precedent remained important means of generating legitimacy for practitioners and practices into the modern period. The contents of these texts have together constructed an understanding of the Ayurvedic system of medicine based on several basic principles. The primary organizing feature is the interconnection of the *dosas* (humours), the *dhatu* (body tissues) and *mala* (waste products). The three *dosas* of the body (wind/*vata*, bile/*pitta* and phlegm/*kapha*) act together with the *dhatu* (chyle, blood, flesh, fat, bone, marrow and semen) and the *mala*. This is called *tridosa-vidya*, the doctrine of the three humours, and underlies theoretical approaches to the body in the canonical Ayurvedic texts. The impetus behind the theory of the *dosas* is the foundation act of attempting to balance them. For instance, a vaid's tasks are divided into two categories linked inherently to the balance of *dosas*: his first recourse against an imbalance of the *dosas* is to calm the *dosas* through dietary or pharmaceutical regimens; his second is to turn to clinical therapies to externally purify the overexerted *dosa* to correct the imbalance.[12] Chapter 15 of the *Susruta Samhita* guide reveals the importance of 'decrease and increase of dosas, dhatus and malas' and teaches students how to recognize the characteristics and functions of these entities through observation of the colour, smell and texture of patient's bodies and excreta, and through an appraisal of the sensations patients can recall.[13] Chapter 4 provides a lesson in 'interpretative discourse', in which vaids are taught about the means through which to distinguish between symptoms, and also how to read and learn. Susruta relays this wisdom by way of inspiration: 'An ass carrying the load of sandalwood feels only the load and not the (fragrance of) sandal, (similarly) those having gone through many scriptures but ignorant of their ideas only carry like an ass.'[14]

The second tenet of Ayurveda is the identification of digestion as a central process of the body. Understood in the literature as 'cooking', the digested food then becomes *dhatu* of the chyle variety. The *pitta* in the body, what allopathy understands as stomach acid, transforms the chyle first into blood, then into flesh and into all of the other forms of *dhatu* until the food finally becomes semen.[15] This also explains the transformation of food into *mala*, including sweat, urine and mucus.[16] The third tenet of Ayurveda explains the physical constitution of the human form, conceptualizing the body as a series of tubes through which the *dosas* flow to the various *dhatu*. Propelling them is *ojas*, energy, which is the source of strength for all bodily functions. Taken together, this understanding of the different components of the body, their constitution and the method by which they digest and excrete food matter together constitute the basis for the Ayurvedic tradition.

Reference is made to these Ayurvedic principles in a variety of contexts, most notably in descriptions of other aspects of physical culture. David G. White, for instance, has studied in detail the link between the body and its maintenance in the Yogic and Tantric tradition.[17] Ayurveda is further shaped by the interventions it makes as a coherent system in discourses that rely only partially on conceptions of physicality and that take into account other factors and phenomena. A key theme of this sort is the relationship between the human body and the environmental landscape. The most pioneering work on this subject remains Francis Zimmerman's *The Jungle and the Aroma of Meats*, in which he

traces the evolution of the Sanskrit conception of *Jangala*, the dry terrain of the Indo-Gangetic plain (as opposed to its corruption in Hindi
as *Jangal* – and the Anglo-Indian term jungle – which refers to the 'tan-
gled thickets' of a luxuriant, marshy terrain), and the ways in which it
is made meaningful to humans.[18] The study began for Zimmerman in
a series of entries in medical treatises – most notably, the *Susruta* and
*Caraka Samhitas* – which listed the properties of meats, dividing animals
into a variety of categories ranging from *jangala* (dry terrain) to *anupa*
(marshy terrain), as well as into herbivores and carnivores, game and
predators, like from like and so on. He argues that the identification
of species of animals discussed in the pages of ancient medical texts
together constitutes a classificatory system that acts as an ancient practice of zoology, a mode of enquiry absent in other Sanskrit writings of
the ancient world.[19] Encompassed in the Ayurvedic text, he argues, are
detailed classifications of geography, vegetation and animal specification. However, what interests Zimmerman about the particular nature of
classification at work here is the way in which the medical literature consistently discusses these three categories in tandem; for instance, 'in a
single landscape, thorny shrubs, bushes, and gazelles may be associated,
or again, a teak forest may shelter lions and antelopes'.[20] This can be further exemplified from what Dalhana says to Dhanwantari that 'satmya
are those which in spite of being naturally contrary in terms of place,
time, race, season, disease, exercise water, day-sleep, rasas etc...do not
afflict', which he glosses as being not only a characteristic of those from
arid land but also characterized by having certain varieties of insects
present.[21]

It is possible for Zimmerman and those after him to locate in the
modern sciences of classification a series of articulations into which to
'subsume the empirical data' of his study, and to privilege a mode of
conceptualizing nature in a particularly local, indigenous and 'ancient'
articulation.[22] Furthermore, the notion of interconnectedness as a mode
of understanding natural habitat brings to light the effect of the medical
approach: nature is primarily understood through a human's encounter
with it; further to this, it is described in the text in the context of illness. A medical understanding of the natural world thus relies upon the
lived environment of the human. In the Indian context, the premise
of Ayurvedic medicine served as a mode for interpreting and ordering
the natural world. Following in this tradition, the Tantric scholar David
G. White has elaborated this notion of interconnectedness by identifying commentaries within the traditional texts of Ayurveda on the effects

of seasonal change, adding time to the space continuum identified by Zimmerman.[23]

The textual basis for understanding the importance of Ayurveda in pre-colonial society tells a significant part of the story but, if discussed in isolation, reifies Vedic culture and Brahmanic customs as representations of normative practice. Exploring the popular practice of Ayurveda throughout its history contextualizes this consolidated tradition within the wider practice of medicine in the northern subcontinent. With its ties not only to Sanskrit learning but also to Indic culture, Ayurveda is considered to be the oldest and most established of the indigenous systems of Indian medicine. However, not all early groups and traditions that influenced and were influenced by Ayurveda can be confined within the mainstream geographical, cultural and religious heritage of modern India. The suggestion that Ayurveda forms part of an isolated Sanskritic cultural and intellectual tradition often relies on the notion of an 'Aryan' migration to the subcontinent and the formation of an Indus Valley Civilization. However, early Vedic accounts make evident the absorption of pre-Aryan urban culture into medical practice. Alongside other continuities, this undermines the argument that there was a distinct Aryan 'invasion' that swept aside earlier groups. Instead scholars of medicine prefer, like many other historians of ancient India, to see emergent Aryan cultures as the coming together of migrant populations with earlier inhabitants of the land, drawing on previous forms of urbanism.[24]

The association between Indic medicine and urban organization is apparently as old as urbanism itself, and certainly as old as Harappan culture. The importance of water in the civic organization of Mohenjo-Daro is discussed in the predominantly philosophical accounts of the early medical practices of the Aryans.[25] Founded in the fifth millennium BCE, Mohenjo-Daro was the first city to be established in the history of human civilization. This focus on drainage and a centralized water distribution system has been taken as evidence of a concern with public health.[26] Personal hygiene and cleanliness, an important aspect of disease prevention in urban societies, have always had a prominent place in Ayurvedic medicine, particularly by comparison with other pre-modern health systems.[27] This should be seen alongside other aspects of medical folklore and 'traditional' practice incorporated into early Ayurvedic texts.[28]

Communities that came into being after the Vedic period have also claimed Ayurveda as a vital part of their tradition and to have aided in its evolution. Various sects that emerged out of the Indic civilization remain part of the broader spectrum of Hinduism and have adopted aspects of Ayurvedic thought as their own. Those who follow the Tantric tradition incorporate aspects of 'Ayurvedic' practice into their rituals, as do yogic practitioners.[29] Most notably, Ayurvedic texts inform aspects of early Buddhist thought on the physical and medical sciences.[30] Kenneth Zysk argues that the early Buddhist pharmacopoeia was derived in part from Ayurvedic classifications of plants, foods and medicine, and that the Vedic codification of this information was instrumental in creating medical tracts for post-nomadic Buddhist communities.[31] Similarly, methods of surgery and the development of surgical tools link the two traditions, as do professional hierarchies.[32] Zysk also stresses the importance of the Buddhist tradition to the spread of Ayurveda outside of the subcontinent, linking Ayurveda to the *Mahayana* tradition of Buddhism and, subsequently, to the foundation of the religious customs of Tibet, Central Asia and China.[33]

A lack of evidence makes it difficult to consider the possible contribution of the medical practice of non-Indic rural communities, particularly those groups that became known as the *adivasis*, or first inhabitants. *Adivasi* connection to and differentiation from the dominant South Asian population has been a subject of much debate for over 200 years.[34] Little is known about modern *adivasi* medicine, and virtually nothing is known of its history. Scholars have largely ignored the implication that if Ayurveda is indeed an Indic, or at least an urban, creation, then its philosophical and intellectual basis should be different from that of the medical traditions of ancient non-urban groups, and they have failed to look for any evidence to the contrary. The delimitation of 'Aryan' culture has marginalized the possible shared heritage of Ayurveda and *adivasi* medicine and has prevented enquiry into the possibility that not just Ayurveda but also adivasi practice could shed light on the 'original' Indian system of medicine.[35]

Though the lack of continuity between the Vedic literature and the canon of Ayurveda was discussed earlier, hymns within these older texts provide some insight into the role of medicine in society. For example, the place of practitioners within the hierarchy of classification was made evident with healers (*bhisaj*) coming between carpenters (*taksan*) and priests (*Brahman*). Moreover, reflecting the biomoral context of medicine, practitioners' work was associated with that of ritualists (*vipra*) and the Brahmans.[36] They were grounded in two separate worlds, at once capable of performing rituals and guarding that knowledge, while at the same time having the skills to produce and distribute remedies, resulting in their spiritual, intellectual, technological and economic involvement in the organization of society. The first account on the role of medicine in the Common Era was recorded by the Chinese traveller Fa Hsien, who attended a medical council at Pataliputra (in contemporary Bihar) in the fourth century and commented on the medical infrastructure of the city[37]:

> The head of the Vaisya (merchant) families in them [all the kingdoms of North India] establish in the cities houses for dispensing charity and medicine. All the poor and destitute in the country, orphans, widowers and childless men, maimed people and cripples, and all who are diseased, go to those houses, and are provided with every kind of help, and doctors examine their diseases. They get the food and medicines which their cases require and are made to feel at ease; and when they are better, they go away of themselves.[38]

Fa Hsien's commentary is particularly important as it derives from a perspective that deviates from the textual and material sources that give less insight into wide-ranging popular practices. Taken together, these observations point to the social engagement with medicine and those who practiced it, vesting them with the authority of skilled tradesmen, but also with that of the morally and spiritually learned.

Romila Thapar argues that the juxtaposition of these two roles is fundamental to an understanding of the status of practitioners in this period: on the one hand, their work with the human body and with animals was sufficiently practical to exclude them from proper Brahmanic status; however, the usefulness of this sort of knowledge resulted in its codification in Sanskrit, which elevated their cultural status, even if this was not reflected in the formal system of social stratification.[39] Thapar also differentiates between practitioners and those who codified this information, rightfully pointing out that devoid of priestly status, most practitioners probably could not read or write, though she does allow for the possibility that some scribes were also practitioners. At the same time, she argues that it was the straddling of both the priestly and more common worlds that allowed Ayurvedic practice to transcend Brahmanic orthodoxy; as medicine was the most profoundly applied physical science and needed to change to reflect the health problems of the day, it could not afford to conform to outdated principles or to be kept in the realm of the sacred.[40]

Thapar has also emphasized the political and social power of codified knowledge in ancient Indian society. For Thapar, standard health practices constituted a body of knowledge about physicality so well-established that insofar as renunciation movements encouraged deviation from those rules, renunciation functioned as a counter-culture.[41] The physical acts of renunciation, including experimentation with hallucinogens, manipulating the functioning of the body (pulse, breathing and heartbeat manipulation), may have been associated with attempts to achieve levitation, invisibility and flight through extreme yogic exercises. Thapar argues that these attempts to deviate from the normative state of physical humanity represented efforts to channel Shamanistic practices that had fallen into disregard, as well as to 'search for a non-Orthodox comprehension of knowledge and in part a means of asserting power through claiming to know the incomprehensible'.[42] The radical character of such attitudes to the body implies a hegemonic character for more conventional health practices.

Ayurveda as a cogent system recognized and identified by name is somewhat harder to trace. One of the rare and earliest exceptions is that of Vatsyayana's *Kamasutra*. Written in the fourth century CE and reframed by sage Yashodara in the thirteenth century, it remains the dominant guide to sex/uality and the body in the classical Indian world. The nineteenth-century rendition of the text, pieced together by the intrepid travellers and proto-sexologists Richard Burton and F.F. Arbuthnot (and an unknown team of pandits), highlighted the most salacious sexual details of the text and the section of sexual positioning (many of which bore no connection to the original text itself), thus interpreting the *Kamasutra* as a virtual guide to sex and sexuality in the subcontinent, devoid of social commentary. Its most recent scholarly translators, Wendy Doniger and Sudhir Kakar, have taken this rendition of the text to begin an investigation into a potted history, and, more broadly, to unpack the categories of 'science' from that of 'art' vis-à-vis sexuality.[43] Theoreticians of sexuality and its history, the most prominent of whom would be Michel Foucault, have constructed the Asian *ars erotica* (arts of eroticism) as being devoid of *sciensis sexualis* (science of sexuality), ostensibly conforming to the Orientalist myth of sexuality in Asia as being something that is understood viscerally, experienced only sensually, bears no rational explanation or foundation and is devoid of the complexities of social or scientific meaning.[44] Instead Doniger and Kakar explore the taxonomies at play in the work to interrogate the integration of 'science' and 'society', identifying in so doing a range of medical information and social analysis that offers a complex reflection on the construction of the South Asian self along the lines of gender, religion, ethnicity, sexuality, region, class and caste.

The relevance of this reworking of sexuality to the history of Ayurveda emerges in the way in which the system is deployed in the text. Doniger and Kakar weave instances of Vatsyayana's mention of a medical tradition referred to as Ayurveda in and out of their introduction, focusing both on remedies borrowed from it for sexual disease and also on examples of Vatsyayana's dissent from the prevailing Ayurvedic cures and his gloss on their interpretation.[45] For instance, Vatsyayana attributes skin care regimes to the knowledge of the *Ayur Veda*, noting that the practice of rubbing one's body with sandalwood is derived from that tradition.[46] The significance of the mention of Ayurveda is twofold. Firstly, it locates medical discourse at the heart of the construction of sexuality, which does away with the binary of art/science and stresses instead on their mutual constitution. Secondly, it identifies Ayurveda as the dominant medical tradition of the time, thus implying that it was entrenched enough as to be able to withstand dissenting views regarding corporal constitution and its maintenance.

The Kama Sutra's competing understandings of embodiment differed from that held within the 'Ayurvedic' tradition, therein testing its veracity as a system. Ayurveda shifted from referring to a series of texts or a collection of ideas to signifying a coherent tradition that was rooted in texts but that was relevant as the dominant mode through which the body was understood. It remained as such throughout the ages, routinely deployed as a marker of a variety of ancient and indigenous knowledge marked against contemporaneous introductions of new and differing systems. In later years, Ayurvedic embodiment was a term posited against the innovations of tantric logics of the body, implying its tenacity as a normative set of ideological concepts that could be pitted against the radical break posed by fringe forms of practice.[47] After the establishment of Islamic courtly life, Ayurveda came to be represented in contrast to the Unani Tibb tradition that travelled to India with the advent of an Islamic political presence from the seventh century.[48] Abu Fazl, the Mughal emperor Akbar's advisor and scribe, mentions Ayurveda at the beginning of the third volume of his impressive *Ain-i-Akbari*, included in a section on the 18 sciences of the Hindu belief system. Ayurveda is described as the 15th science, composed of 'the science of anatomy, hygiene, nosology and therapeutics... taken from the first Veda'.[49]

In a rather minor aside contained within a broader argument about ontologies of good health in South Asian medical contexts, the anthropologist Joseph Alter attempted to explain varieties of Ayurvedic 'modernity'.[50] In defining the subject matter to be analysed and considered within the scope of his ontology, Alter claimed that he was emphasizing a division between two meanings of Ayurveda:

> It should be clear but needs to be emphasised that I am making a sharp distinction between Ayurvedic theory as represented in the canonical literature and in contemporary technical, popular and academic interpretations of that literature, on the one hand, and applied Ayurveda as it is practiced in hospitals, clinics, and research institutes in South Asia and elsewhere on the other.[51]

Alter comments further on this distinction in a footnote, mentioning that he was using the works of Caraka and Susruta in contemporary translation, without paying heed to the criticisms that Sanskritists might lodge at him about the quality of the texts he chose. 'I use these texts rather than relying only on the unadulterated "authority of the scriptures"', Alter argued, 'to make the point that an Ayurvedic theory of metaphysical fitness is as "modern" as, for example, the prescription for shingles written out by a physician working in an Ayurvedic clinic in contemporary New Delhi, Bombay, or Madras.'[52]

Alter's justification of his methodology brings two major insights of relevance to the task of historicizing Ayurveda. Firstly, he asserts that there is a division between text and practice that conforms to the tension between these two poles of tradition that we have visited in different periods. Secondly, he insists that the consistent marker of the tradition is precisely this tension and *not* the indicators more casually deployed to chart time or chronology. Textuality, and not a specific text, is a fundamental principle of meaning in Ayurveda, so is practice, though not any specific technique. Ayurveda is thus fundamentally organized around its dual existence, as both theory and practice.

Taken together, the deployment of Ayurveda as medical signifier pointed to a singular, historicized outcome that aligned Ayurveda with an ancient past. Ayurveda served as a catchall category for vaguely Hindu, thoroughly indigenous and mostly unhistoricized sets of practices or ideas that were pre-extant in the subcontinent before the advent of Islamic and allopathic medicine. Within these deployments there is no possibility of accounting for the extra-textual evolution of a cohesive (or messy, for that matter) set of ideas or innovations reliant upon the logic of Ayurveda. The vast chasm between textual significance and the pragmatics of practice is insurmountable in this casual and yet profoundly decisive rhetoric that consistently referenced a slice of the past to account for its entirety. Yet, at the same time, the occlusion of practice and lived experience was perhaps not crucial to the Ayurveda-as-reference-point at work in these moments. Perhaps it is to other moments, to other conversations, that historians must turn to evolve our understanding of Ayurveda's complex histories.

Historians of India's medical pasts have framed this question of representation by posing larger ones about the production of medical systems in the subcontinent. Was Ayurveda a historical system of medicine? Can the idea of a system of medicine truly account for the complexities of a medical tradition that poses ambivalences towards textuality? How would the regional and linguistic divergences that alter Ayurveda's manifestation measure up to a model that holds a unilateral truth about the body at its centre? And what of the slippery role of religion in all of this?

Historians are split on the issue. A pioneering 1976 study by Charles Leslie called for an upheaval of the conceptualization of traditional Asian medical systems, suggesting a move away from focusing on their radical differentiation from biomedical systems, and instead drawing to attention the particular historical and cultural processes through which they were formed.[53] Leslie's study undertook a comparative view of medical systems in India (Ayurveda), China Traditional Chinese Medicine (TCM) and the Middle East (Unani Tibb), in order to examine the ways in which norms of practice, designation of authority and expertise, and adherence to key texts underlay each tradition. Leslie's intention was not to compare these traditions to their Western counterparts but rather to argue for their individualized histories of internal coherence and cultural relevance.[54] While the particularly modern evolution of Ayurveda as cultural, political and social practice informs the basis for my study, Leslie and others like him would take a similar approach to representations of the ancient past. Ayurveda to Leslie is thus most accurately represented as a system that evolved over a long period of time, and its history is illuminated through its relevance to various themes and genres of the South Asian past. In constructing a genealogy of Ayurveda, its historical and cultural relevance can together inform the intellectual coherence and social resonance of the tradition.

While Leslie's argument is a compelling one, historians of modern India have noted the limitations of the systems approach, arguing that it follows too closely the early Orientalist model of trying to define and taxonomize the scientific and intellectual worlds of Indian difference. The anthropologist Jean Langford, in an interdisciplinary study of contemporary Ayurvedic practice in the twentieth century, warns against the seduction of 'systems', urging scholars to 'resist the temptation to fix Ayurveda into a discourse of order as a classical medicine operating according to a strict logic'.[55] David Arnold sees a range of difficulties with the systems approach, beginning with the Orientalist strategies used to systematize scientific knowledge and following a Linnean model along an improvised historical axis, so that Ancient Hindu Medicine stands in opposition to Medieval Islamic Medicine, which will be overtaken by Modern Biomedicine.[56] This is further impacted by the reductive attempts to determine the notion of the 'scientific' within traditions with complex and ambivalent relationships to the rational categories of science. Equally troubling for Arnold is the attempt to argue for an internal coherence, an approach that insists upon a singularity anathema to the great regional, ethnic and linguistic differences at play and that inspire different manifestations of similar principles – along with outright contradictory stances upon the most basic claims. In a clever turn, Projit Mukharji turns the systems metaphor on its head by applying it to Western medicine in the subcontinent, examining the absorption of biomedicine into Bengali medical life, and exploring the ways in which the new category of 'daktari' came to present a challenge to the notion of a coherent allopathic system through its fluid adoption of indigenous and biomedical principles into moments of practice.[57]

Guy Attewell's study of the reconfiguring of Unani Tibb in the nineteenth century further problematizes the neat ways in which the word system 'consolidates the impression of continuity, connoting internal coherence, discreteness, completeness, homogeneity'.[58] Attewell instead locates the process of system-making within colonial-era attempts to demarcate and represent knowledge and practice as a coherent whole, arguing that the Tibb-i-Unani came to occupy the place of a medical system through a complicated series of negotiations with the medical modernities introduced by the colonial state, and reimagined by indigenous actors. As we shall see, similar claims can be made about the trajectories of Ayurveda in modernity, especially vis-à-vis the question of politics. While Ayurveda is crucial to both arguments for and against the use of the term system, the conceptual arguments are somewhat limited by the pragmatic of its use in twentieth-century India.

Scholars and studies of practice and performance are justifiably hesitant of the systems moniker, and have reframed the ways in which the indigenous medical traditions of the subcontinent can be theorized according to alternative logics and rationales that can account for their interdisciplinary accommodations. Rather than pursuing the systems debate further, we will move beyond it to explore the ways in which *systematization* was imposed upon Ayurveda as a way of easing its coherent entry into formal politics. While Orientalist at its foundation and wholly (and perhaps purposefully) unaccommodating of the complexity of the intricacies of Indian embodiment, an overview of the systematization of Indigenous Medicine reveals the pragmatics of marginalization and cooptation of these traditions within (and beyond) the reach of the colonial state. We will explore the ways in which practitioners, technologies and logics of embodiment were disciplined by the biopolitics of empire. Moreover, we will explore the range of ambivalences, refutations, capitulations and assimilations that together forge a more complex lens through which to examine the mechanics of systemization within the structures of health governance.

### Identifying the biomoral in theory and in practice

While the categorization of Ayurveda as a system of medicine is cause for some debate amongst scholars, the conception of Ayurveda as a *biomoral* tradition acts as salve. The term has become a commonplace in discussions of embodied practices that incorporate a sense of morality into their logic of practice. With regard to medicine, the biomoral often describes the exceptionalism of a local approach to conceptualizing the body and its functioning. It is made to represent the dominant systems of regional, culturally specific scientific systems that are often articulated in part through local traditions or belief systems concerning the body. This notion of the biomoral is certainly an adequate mode for analysing the effects of cultural encounter on the rigidification of medical systems, but can easily fall prey to a worldview that might privilege allopathic medicine as the global norm against which regional, local and 'indigenous' systems of medicine might be pitted and found lacking.[59] This approach also assumes that biomedicine is divorced from the social, cultural, economic and political context in which it emerged.

The articulation of the biomoral as a theoretical model through which culture and society in the South Asian context can be explored has been most thoroughly covered in anthropological writings on the body and its place in South Asian society. Most prominent, even 30 years and several ideological shifts after its inception, is McKim Marriott's idea of the 'biomoral logic' of interaction across caste and class lines in South Asian society.[60] Marriott's work was intended as a direct contestation of the neat taxonomies of hierarchy envisioned earlier by Louis Dumont's classic and infamous text *Homo Hierarchicus*, which employed a structuralist anthropological approach to understanding caste as a hierarchical system of symbolic purity. Marriott's challenge to social scientists of his day was to identify a series of categories of meaning within the South Asian context that might undo those conceptualized in the West and imposed upon non-Western societies. His idea was to '[construct] an alternative general theoretical system for the social sciences of a non-Western civilisation, using that civilisation's own categories'.[61] This notion of South Asian personhood is made evident in studies of transaction and gifting undertaken by Marriott and others, in which 'they must also give out from themselves particles of their own coded substances-essences, residues, or other active influences – that may then reproduce in others something of the nature of the persons in whom they have originated'.[62]

Marriott's model argues that South Asians fundamentally conceptualize their embodied selves as being monistic, 'dividual' assemblages made up of both bio-genetic substance and moral code. Considered in light of its theoretical genealogy, the notion of the biomoral has evolved over the past four decades of Indian anthropology and has raised new understandings about the connection between the bio-genetic codes and moral frameworks that together determine human behaviour. The biomoral for Marriott literally refers to the confluence of morality written on to the biological form of the Hindu and the ways in which it extends through the giving and receiving of gifts. It is the purity assigned to the biological form of the moral being, represented in caste status, which is affected through giving. The monistic, dividual model he presents allows for an envisioning of all forms as fundamentally connected through the necessary state of overflow that complicates the rigidities of purity, stratification and 'boundary-oriented' theories through which South Asian culture has been previously observed.[63]

Most importantly, Marriott bolsters his insistence on the necessity of over/flow as a state of being in South Asian culture in a reading of traditional Ayurvedic practices, as represented in the textual tradition. The flows of *ojas* (energy), the practice of 'cooking', the paucity of the borders between the imbibed substance and its excretion in some other form together inform the model of complex, unbounded flows reflected in morally resonant social and cultural mores around gifting. If the body cannot be the sole container for the substances that flow throughout it, then the moral structures that emphasize division must rationalize the omnipresence of flows that may threaten the structural hierarchies that divide individuals. The moral code according to which South Asians interact is re-inscribed with the pragmatism of biological flows. At the same time, the biological state is vested with the weight of moral value that its functioning might support or threaten. As Jonathan Parry argued two decades after the introduction of Marriott's model of the biomoral, 'substance determines conduct; conduct modifies substance'.[64] Far from being a vague practice of unclear proportions, Marriott and his colleagues trace the myriad crossover between the imbibing of food or the maintenance of the body and its supposed effect upon character. An individual's character, for instance, '[is] thought to be altered by changes in the person's body that result from eating certain foods, engaging in certain kinds of sexual intercourse, undergoing certain ceremonies, or falling under certain other kinds of influences'.[65]

The root of these characterizations and their 'known' connection to different bodily practices are informed by notions of ancient textuality, bringing the foundational texts of Ayurveda to the forefront of modern practice. The anthropological approach to the biomoral initiated by Marriott and Inden (and continued into the classical work of Jonathan Parry and very recent work by Lawrence Cohen, which deal, quite literally, with the preservation of life and the eventuality of death in Hindu South Asia) identifies the *Susruta* and *Caraka Samhitas* as the most appropriate sources when questions of 'evidence' and 'precedence' arise. For Marriott and his generation, the question of textuality was incorporated into the approach without the caveat of a historical problematic: Ayurvedic flows as identified in Indological medical works served as foundational points of reference in the arguments they expounded on the fallacy of rigid individualism. Textual evidence created an ancient precedence for understanding the ways in which South Asians self-conceptualized their dividualism and monism as made manifest in acts of giving. Jonathan Parry, however, problematizes the supposed coherence of metaphysical identification that Marriott insists upon. While upholding Marriott's model as a viable critique of Dumont's dualism, and while employing the model of the substance-code connection therein as a foundation stone of his own investigation, Parry allows for 'a robust and stable sense of self', to which most of his participants gave voice despite their monistic, dividual connectedness. Parry suggests that the model that Marriott and colleagues have created is somewhat overdrawn, and, by way of intervention, poses this question to them: 'how indeed can anybody ever decide with whom, and on what terms, to interact?'[66] More recently, Jacob Copeman's elegant study of blood donation in contemporary India, which draws heavily from the anthropology of the gift, addresses the question of intention around instances of blood donation across lines of caste, class, community and religion, therein pushing this question well beyond the bounds of notions of ritual purity for caste Hindus.[67]

Marriott's conception of the biomoral relies upon a model where the foundational texts of Ayurveda remain at the centre of their explorations of body culture, thus reifying the 'ancient' texts as the basis for contemporary knowledge production. The relevance of literature is always punctuated in the South Asian context with the question of literacy, inspiring scholars to frame the text more as cultural object than as consumed good.[68] The Ayurvedic text, in all of the instances described, fits this model: more than passing down a literal truth about the body, it instead provides a framework to discuss the pervasiveness of ancient knowledge in various historical moments. The designation of the biomoral in its more literal interpretation designates the symbiosis of the abstract and the pragmatic by insisting that medicine, in some cultures, is not only about 'scientific' reactions. In addition, it inherently employs the notion of 'ancientness' to also contest the biomedical insistence on 'modernity'. The biomoral as conceptual mode takes this idea further: instead of mediating these relationships between the ancient/traditional and the modern through the veracity of the text, it instead focuses on new categories of meaning that take into account the application of this discourse of textuality as a means of shaping culture.

More recent anthropological approaches have come to reframe the question of the biomoral along new axes of signification. Langford's avowal of the biomoral is conceptualized loosely in response to her notes on the seduction of systems. For her, the biomoral occupies the gap left by the limitations of allopathic logics of embodiment, made manifest in the organization of texts or the norms and performance of practice. For Langford, the biomoral can accommodate the logics of mapping and organization that follow a counter-impetus for ordering, where the restoration of illness, framed as a liminal time in which 'ordinary social meanings are interrupted by unintelligible pain and incapacity', is achieved through a realignment of the social, cosmic and somatic order.[69] The pragmatics of organization is also inflected by other influences, for instance the ordering of topics in terms of *namamala* (garlands of names) rather than taxonomic hierarchies.[70] Working in a different context but along similar lines, Joseph Alter's pioneering analysis of Gandhi's 'biomoral' self-disciplining draws on a similar reorienting of embodied signifiers along new axes of meaning.[71] Alter's work on Gandhi reveals the ways in which his embodied practices like celibacy, fasting, cotton-spinning and vegetarianism always carried with them a firm rooting in anti-colonial rhetoric, referencing a very political nationalism directly tied to the state. Gandhi wrote several guides on the body and its maintenance, in which he advocated a return to 'natural', indigenous systems of caring for it. Both Langford and Alter leave room in their analysis for the permutation of a specific body politic as aggregate to the biomoral.

Finally, Lawrence Cohen's groundbreaking work on the organ trade in South India furthers the political possibilities of the biomoral beyond any other reckoning of it. Working through the complexities of kidney transplants through a series of vignettes ranging from ethnographic interviews to popular filmic representations of transplantation, Lawrence moves Marriott's conception of biomoral transactions across caste, gender and generation into the framework of the Nehruvian developmentalist state.[72] The understanding remains at work here, filtered through Donna Haraway's conception of coding, in which science desires 'the translation of the world into a problem of coding...where heterogeneity can be submitted to disassembly, reassembly, investment, and exchange'.[73] Cohen reads these two varieties of coding in tandem against the backdrop of a viable political ideology, which forces the abstraction of Mariott's notion of coding to adapt to the fluid, unified code of Haraway's as a metaphor for postcolonial, developmentalist ideologies of Nehruvian nationalism. However, rather than seeing transplant blood organs as absent of code, Cohen instead proposes that blood and organs are reinvested with reformist claims of ancient, pre-Brahmanic liberal forms of unity in the Mother India.[74]

Cohen then considers the work of suppression in this model, using Agamben's distinction between the idea of bare life (*zoe*) and political and human life (*bios*) to delve into the larger issue of sacrifice at play in the discourse of donation. Where state narratives of transplantation focus on the health of the recipient, there is a concurrent abandonment of the donor, especially under the messy circumstances of unregulated organ donation. Agamben, drawing on Aristotle, understands *bios* as the legally protected human life of sentient beings with the moral and political world of the *polis*, while *zoe* remains outside of the spectre of the law but still under the protection of the sovereign. Agamben understands this distinction to be most relevantly reflected in the Roman legal principle of *homo sacer* – a person whose life is placed in the space of sovereign exception, like a brain-dead person or a concentration camp internee, whose death would not be considered either murder or sacrifice.[75] Cohen argues that the kidney is *zoe*, as it exists outside of the realm of morality, politics or social flow; in essence, the sacrifice around its donation is unrecognized. Thus, a biomorality of 'inassimilable difference' (caste, religion) is abandoned as the body is reframed as a bag of organs from which individual components can be severed or replenished.

Cohen's shift from the logic of the biomoral to a framework of biopolitics signifies a break from prior readings by insisting upon a return to the fundamental political questions of the day as a crucial component of moral or ethical reasoning. As we will see throughout this study, the delineation of Ayurveda as a moral, ethical or spiritual practice with a biological component was inherently tied to the politics of nation-building, colonial resistance and state-building. Rather than precluding its separation from the biopolitics of late colonialism, Ayurveda's biomorality ushered in and legitimized notions of the authentic, indigenous body, an exemplar of the modern Indian citizen, and in opposition to 'foreign' (read Muslim) trajectories embodiment. As we shall see, the biomoral and the biopolitical together shape a genealogy of Ayurveda's induction into the pragmatics of late colonial health governance.

### The biomoral in action: Ayurveda's entry into modernity

In more pragmatic terms, Ayurveda's biomorality was shaped by its inherent reference to an ahistorical, amorphous notion of ancientness that it was meant to represent in modernity. In essence, the characterization of its 'ancient' origins of its bio-content allowed for the system to become a framework for debating the tension between traditionally indigenous and contemporaneously foreign morality. Ironically, Ayurveda's elusive biomorality is invoked by biomedical dissenters, who laud the system for its fluid adherence to the rigors of evidence-based principles, and who praise its alleged incorporation of esoteric, spiritual or 'holistic' health principles.[76] On both sides of the biomedical divide, it is agreed that Ayurveda's moral features outstrip its biological components. This reading of Ayurveda lends more insight into Orientalist fantasies and constructions of Eastern embodiment than to the pragmatics of framing Ayurveda within the moral structures of embodiment in South Asia. However, it also reveals the importance of the concept of the biomoral – from its theoretically precise to its 'not quite science, not just religion' mode – to Ayurveda's entrance into the realm of the modern political. The biomoral fundamentally mediates Ayurveda's entry into modernity. In this section, the three realms of the biomoral in practice – the early Orientalist, the Raj's reluctant pragmatism and the anti-colonial biomoral – are explored to consider the applied effects of the biomoral in practice.

In the eighteenth century, Ayurveda was lauded by the noted Orientalist Sir William Jones as a key area of Hindu philosophy and history, explaining that 'Ayurveda was delivered to man by Brahma, Indra, Dhanwantari and five other Deities; and comprises the theory of Disorders and Medicines, with the practical methods of curing Diseases'.[77] This categorization of the divine roots of Ayurveda firmly sealed its position within a framework constituted of belief, and likely devoid of fact – and therefore quite at odds with enlightenment practices around science and medicine. At the same time, the texts designated within the Ayurvedic tradition – namely the sixth-century *Susruta* and *Caraka Samhitas* – did document a physiological and diagnostic logic in line with certain principles of both Hippocratic and Galenic medicine, and did constitute some truths about the body and its treatment accepted within biomedicine.

The basis for comparison was taken up in early liberal practices around education and knowledge production in the subcontinent, which saw the development of hybrid educational institutions where a variety of systems of knowledge were taught concurrently to fulfil the cosmopolitan interest in mapping out global knowledge systems.[78] In 1822, the idea of a Native Medical institution, where Indian practitioners could be trained to perform the sub-duties of European biomedical doctors, was proposed to the government. The duties that could be undertaken by these practitioners would be those of the variety 'that no Medical Gentleman properly qualified would undertake them except on the condition of being handsomely rewarded for his labours'.[79] The idea was to build upon the uneven expertise of Indians already culturally recognizable as doctors by offering them free training if they remained in the service of the government for 15 years; the cost to the government would be 'trifling' as compared to the salaries of general surgeons or other Anglo-Indian practitioners within the medical service. At the same time, the 15-year clause prevented them from practising privately, and hence providing competition for other biomedical practitioners. The lectures would be given in the vernacular languages by an instructor 'with a considerable acquaintance with the written and colloquial languages of the country... [and who is] capable of reading the Native Systems of Medicine and of discussing and conversing with his pupils on all ordinary subjects of medical science in intelligible if not in accurate terms'.[80] Though this was clearly an institution created for the benefit of the colony, it reflected the hybrid spirit of the early nineteenth century, where the veneer of cultural 'exchange' between Eastern and Western knowledge about medicine, disease and the body could legitimately characterize the venture. At the very least, it entertained the notion that indigenous logics of medicine and the body, steeped as they might be in the vagaries of religion, proved useful (and, at times, crucial) to the project of medical planning in the Indian colony.

The institutional pragmatics of inquisitive and expansive liberal interest in multiple knowledge systems collided with a shifting imperial politic that privileged only one kind of learning for both Indians and Europeans alike, resulting in the Anglicization of education after 1835.[81] The Native Medical Institution shut its doors in 1835, and similar classes in the vernacular languages and indigenous cultures of medicine at the Calcutta Madrassah and the Sanskrit College, the two other major educational institutions in the city, were abolished. The Calcutta Medical College was founded in their stead; as David Arnold has made colourfully evident, the crowning act at the Hindu College was when a Brahmin instructor led the dissection of a cadaver, undeniably representative of pollution in its most vivid form.[82] More intuitively, the focus on the language of instruction and education, here, provides the key to the moral question at hand: the dominance of a certain variety of morality, bound up in Sanskrit texts and Hindu ritual, proved anathema to a shifting colonial context in which knowledge needed to be transmitted in a solely Anglicized idiom. Ayurveda was simply dismissed from the realms of formal education because it could not rise to the challenge posed by Anglicization – it was too mired in the particulars of the Sanskrit language, as well as lacking in an evidenced-based logical underpinning, to meet the criteria for inclusion in the modern, English language curricula of colonial medical education.

The formalization and Anglicization of medical education was at best met with ambivalence (and, more pragmatically, a total unawareness) by indigenous medical practitioners who had no expectation of working within the framework of any state, and who likely lost very little of their business of healing to biomedical institutions. This exclusion did not affect the ongoing practice of Ayurvedic or other indigenous medical systems, as medicine was traditionally practiced on the local, intimate scale of the village or the family and had historically little to do with larger state structures. In fact, Unani medicine enjoyed a period of revamp and restructuring in the nineteenth century, particularly with the onset of Urdu publishing in the 1860s, and with the formation of gentlemanly societies of Unani doctors from the 1880s on.[83] Many Hakims responded to the new medical institutional changes by altering their standards of practice to conform to the vision of medical modernity promoted by the colonial government. Seema Alavi has identified methods that Hakims in the UP adopted to re-establish their profession and professionalism after the destruction of elite patronage, characterized by the regrouping of networks around new ideas about the role of the practitioner.[84] This marked a movement away from the authority of family-run practices and educational institutions, and also away from the authority of the Arabic manuscript. Instead, the new Hakim gained legitimacy in the eyes of the public by 'knowing the correct *akhlaq* (norms of behaviour), understanding religion and astrology and their influence on health, and appreciating the personal relationship between the hakim and his patients'.[85] These new ideas were introduced to the public through texts and articles on Unani that circulated in the public sphere, and resulted, Alavi argues, in public debates between the family-centred Unani and the new Hakim. This strategy reinforced the inherently Islamic aspect of the Unani tradition and replaced the old patronage structure that had allowed Hindus to be counted amongst the ranks of Hakims.

Rather conversely, the Indian Medical Service (IMS) lay at the mercy of indigenous medical practitioners. The aim of the IMS in the nineteenth century was to offset any disease or medicine-based causes for the disruption of imperial governance or economy. The primary concern of the IMS – and the impetus for various early health acts passed by the East India Company – was the health of the army, under attack from venereal disease and more generalized occupational hazards. Eventually, health policy translated into the protection of the Indian masses, whose health and living conditions were understood as a political issue. The ill-health of the masses was bad for the imperial economy, due to the loss of labour and the cost of healing; at the same time there was substantial fear that the ravages of disease might lead to revolt. The IMS' solution was to inoculate the population preventatively and to treat diseased subjects during times of famine or epidemic.[86]

The moral slippage between colonial economic venture and agricultural disaster has been well highlighted by scholars who have characterized the relationship as one of 'late Victorian holocaust', to more benign readings of the connection between the two.[87] From a history of medicine perspective, what can be gleaned from readings of health disasters in the nineteenth and early twentieth centuries is the way in which they brought biomedicine to the forefront of everyday life of affected populations. While a more complicated biopolitic was certainly at work changing the lived experience of physicality in the subcontinent – made acutely evident through the institutional disciplining of the body – the advent of biomedicine was still something that the majority of the population, especially those based in rural areas, could mostly ignore.[88] Vaids and Hakims, along with regional and local variations on the figure of the health practitioner, were predominantly responsible for doling out the implements of healing.

During times of crisis, it was precisely these practitioners to whom the Raj would turn to. The recent scholarship on imperial public health has worked to think through the dominance of the Raj in implementing medical policy, especially vis-à-vis vaccination and other preventative campaigns. Biomedical techniques and technologies were sometimes accepted without resistance, and other times resisted violently; however, in many cases, local health practitioners were called upon to aid in the distribution and dissemination of local campaigns, a practice that continued well into the twentieth century, in episodes we will encounter in the following chapters on events in the early twentieth century.[89]

It is in this instance that the second facet of the biomoral can be deduced: local indigenous practitioners lent their moral authority as trusted healers to the campaigns of the imperial government. From the onset of imperial interest and concern with Ayurveda to its eventual dismissal and erasure from imperial life, Ayurveda had come to be constructed in both imperial discourse as illiberal, unscientific and ahistorical: writ large, Ayurveda was nobly ancient but irrelevant. During public health campaigns, however, when IMS officials were forced to rely upon Ayurvedic practitioners to implement medical policy on the ground, Ayurveda, through its newly useful practitioners, was assigned a new set of characteristics: trustworthy, familiar, locally relevant and soundly reliable.

In essence, in this period, the biomoral imperative came to be framed through the racialized difference of the foreign European and the Indian indigenous. Where Ayurvedic tradition had historically accounted for only a loose connection across regions – in practice, regional difference was the focus of debates – Ayurveda was now held up as the morally appropriate vehicle for the treatment of Indian bodies. The assimilation of Ayurveda into practice on the ground, and the subsequent Ayurvedic adoption of certain Western techniques and technologies, complicated but also formally articulated a format for medical integration. Western biomedicine could be applied if introduced to the native populations within an Ayurvedic framework of disease and treatment, fundamentally reassigning a colonial moral framework upon biological principles. It was, in this context, morally appropriate based on the perception of what it could secure, enacted through the trusted figure of the vaid or Hakim who administered the new technology to the sceptical masses. Hinged upon their relationship to the authentically indigenous, the biomoral imperative linked race and place to a historic construction of the Indian body.

This association between health, race and the body in colonial India came to define the unfolding of Ayurveda in the twentieth century and also to articulate the larger project of resistance during the early days of nationalist organizing. The third way of thinking about the biomoral is through the enactment of embodied activism, in which the body became a key site of resistance to colonial rule. The nexus of swadeshi, satyagraha and Gandhian brahmachari as anti-colonial strategies brought the body to the centre of the anti-colonial struggle, tying the moral to the larger nationalistic cause of independence from British rule. The sceptical trust in the embodied biomedical technologies – made possible through the involvement of the complicit vaid on their implementation – was transformed in the twentieth century into a discursive rejection of all things foreign and adoption of the morally appropriate authentic indigenous.

The swadeshi movement of the early twentieth century was the first movement to introduce the idea of universal Indian body as a political strategy. Based on the concept of economic resistance, Indian consumers were encouraged not to purchase foreign goods, and to instead buy swadeshi – literally, of the land. While the logic of the campaign was conceived to cripple the colonial economy, the more powerful political outcome of the campaign was the image of a nation in 'home'-made goods, most notably the powerful visual of locally spun cotton cloth called *khadi*. As Lisa Trivedi has made evident, the use of *khadi* in hats, flags, saris and kurtas introduced a visual vocabulary of the national, rooted in a material culture of nationalism, which could be deployed amongst the literate urban and illiterate rural milieus of the burgeoning Indian nation.[90] C.A. Bayly and Lucy Norris have argued that the wearing of *khadi* elicits a multitude of personal meanings linked intimately with biological understandings of the interaction of cloth with skin but that also resonates with larger social, cultural and political projects.[91]

The most popular figure to medical historians interested in issues of nationalism and embodiment is, of course, M.K. Gandhi. Joseph Alter's work on Gandhi's 'biomoral' self-disciplining as a direct mode of nationalist resistance has initiated a discussion about the embodiment of nationalism through similarly marked acts of resistance.[92] First and foremost amongst embodied practices was the notion of *satyagraha*, or civil disobedience, in which *satyagrahis* were asked to reject violent measures of resistance in favour of peaceful, non-violent ones. This particular ethic of non-violent protest and resistance wed Christian and ancient Jain teachings together, cited by Gandhi as originating both in the teachings and examples of Jesus Christ, and in the historic practices of *ahimsa*.[93] Beyond satyagraha, Gandhi's advocacy of embodied practices like celibacy, fasting, cotton-spinning and vegetarianism always carried with them a firm rooting in anti-colonial resistance, referencing a very political nationalism directly tied to the state. At the same time, he put himself through extreme experiments in embodied resistance that worked to rejuvenate his commitment to living a life of morals. For instance, Gandhi was known to take young women into his bed in order to test and conquer his desire for them, with allegedly mixed results, all in the name of his commitment to achieving *brahmachari*.[94] In satyagrahic moments of protest, the body was literally the site of resistance and of moral provenance – a *satyagrahi*'s embodied non-compliance, either theoretical or very real, was itself a moral critique of the cycle of violence that colonialism had wrought upon the subcontinent.

*   *   *   *

In essence, the weaving of the bio and the moral was, fundamentally, a reflection of the way in which the body was deployed in larger social and political contexts. Ayurveda's entry in the realm of modern medical governance resulted in the colonial state condemning its substantive logics and the practices of its practitioners, while often relying on the social and cultural capital of the system and its doctors to implement public health policy. In the early stirrings of nationalist organizing, the Indian body was reprised as a site of resistance, investing the moral with a new ethical imperative, engaging the bio through acts of dissent and protest. Ultimately, the deployment of tradition as a pole against which modernity could be measured created the moral conditions for the full participation of indigenous subjects in civic life.

A survey of Ayurveda's long history reveals a consistent tension between the realm of the conceptual and that of the pragmatic; both come together, as we've seen, within the space of the political. Any attempt to historicize Ayurveda must therefore begin with a consideration of the extant tension that gives voice to the political particulars of the different historical epochs in which Ayurveda was both crafted and deployed. The systems of knowledge about Ayurvedic medical theory were produced in Sanskrit by Brahman Pandits; the practice of Ayurvedic medical techniques by practitioners outside of the high priestly tradition allowed for Ayurveda to be situated in the everyday life of the community. The tension between these two spheres of meaning resulted in the employment of a wide range of ideas under the rubric 'Ayurveda':

the *Susruta* and *Caraka Samhitas*, now considered to be the foundational texts of Ayurvedic medical theory, were no more representative of the tradition than were the broths sold in the medieval marketplace as cures for fever. While social and cultural codes of meaning determined the value of these two varieties of knowledge, lauding different sorts of social and cultural capital upon their disseminators, both existed equally under the wide banner of Ayurvedic medicine.

At the same time, the intervention posed by colonial rule in the subcontinent became an important part of the way in which meaning was created about Ayurveda. The early colonial history of science and medicine relied heavily on the work of Orientalists who collected specific varieties of information about the subcontinent and attempted to translate it both into English and into the Indian contribution to the history of human civilization they were hoping to piece together. For Orientalists, and later for the colonial state that relied on this sort of knowledge to create social and political policy, Ayurveda represented an ancient Hindu medical tradition based on Hippocratic principles that had sustained Hindus through centuries of Muslim rule. Ayurveda, therefore, was implicated in justifications for colonial intervention, and also in the wider project of vilifying Islamic intellectual traditions in the early colonial period. Ayurveda's indigeneity was reinscribed as being inherently textual, whereas Ayurvedic techniques were thought to have been corrupted by centuries of Islamic rule. As we shall see in the following chapters, the lasting implications of this communalist reading had far-reaching consequences, as Ayurveda was employed to invoke the authentic Indian indigenous.

# 2
# Situating Ayurveda in Modernity, 1900–1919

In this chapter, I consider the processes by which indigenous medicine was marginalized and then resurrected by the colonial state from the late nineteenth century to the first decades of the twentieth. In these discussions, the romanticization/vilification of the indigenous medical systems finally came to an end and was replaced by the beginnings of serious, systemic enquiry into the state of medical life in the subcontinent. The Medical Department found that myriad networks, relationships and experiences of Public Health policy were heavily reliant upon indigenous cultures of physicality, practices of extra-biomedical medicine and pre-colonial patterns of medical consumption. Ayurveda in the nineteenth century had focused primarily on scientific texts and their (re)production and had been more concerned with pandits than with practitioners; however, the Medical Department's new, more pragmatic view of medical systems in action forced the focus to turn to the interactions between doctors and patients. Situating practitioners at the centre of discourse marked a profound shift in the way in which Ayurveda was conceptualized, both by the state and by the new practitioners.

A series of episodes forced the Government of India (GOI) to address the indigenous medical systems head-on in the early decades of the twentieth century. The pattern of devaluing lived practice – and the simultaneous romanticization of an irrelevant past – was problematized, albeit with great ambivalence and much hesitation, as other pressing policy developments called for a meaningful engagement with indigenous medicine. Firstly, the Indian Medical Service's (IMS) late nineteenth-century enquiries into the possibility of an indigenous drugs industry, framed mostly to aid in the newly established field of bacteriology, resulted in a survey of medical knowledge and those who held it. Later on, a request placed by the Hakim Ajmal Khan (and backed by a senior IMS member) to move his practice from Old to New Delhi forced the GOI to draw a line between charitable medical endeavours and the practice of 'unscientific' medicine. Finally, the introduction of legislation to prevent 'quack' practitioners from using their illegally obtained degrees makes evident the extent to which non-biomedical but equally legitimate practitioners were an integral part of the Public Health venture in the subcontinent.

These episodes were fractured and disconnected, reflecting momentary or local concerns with the specific details of medical life, and often addressed the Indigenous Systems of Medicine in rather roundabout ways. However, what ultimately connects these three moments in the early twentieth-century history of indigenous medicine are the questions posed in each instance about the nature of the indigenous medical systems and their practice: What was Ayurveda? What was Unani? Who practiced these systems, who used them, and how? The government was forced to pull apart the categories of knowledge production and implementation, which resulted in a serious discussion of practitioners, patients and consumers. While the impetus for major reform and serious engagement would be ushered in post-war with the advent of a dyarchic structure of governance, the episodic interventions in medical planning laid the groundwork for a restructuring of the indigenous medical systems with a view to their incorporation into a modern structure of governance.

## Making medicine colonial

The colonial state's concern with the health of the native population was, as Sarah Hodges has termed it, both *object* and *mode* of governance: the *object* of policy aimed at improving the health of the people but also the *mode* through which the regulation of daily medical practices could be transformed.[1] The reach of biopolitics in the early imperial period was at once incredibly intimate but also rather selective. Certain daily practices involving the body became subject to the legislative arm of the state and clearly had the subsequent effect of encouraging a disciplining of the self as a subject of empire. Crucial to this endeavour was the delineation of the population at hand through various means of 'measuring' the health of its members. The collection of information about the population was a strategy of use to the state beyond the IMS, and was enacted through censuses, surveillance and data collection that contributed to the government's attempt to 'know' the shape, breadth and strength of the population. However, as Arjun Appadurai and others have made evident, the practice of collection itself was often experienced as an intervention. The imposition of rigid colonial categories of personal and group identification ran contrary to the complex processes of claiming identity and negated the possibility of intersectional identities in favour of a primary designation, such as 'Brahmin' or 'Tribal'.[2] The rise of public health as a preoccupation of the state was a direct outcome of the post-Mutiny government's concern with the security of its position. The GOI adopted new techniques and new strategies to identify, discipline and order the threat of illness and disease to the maintenance of imperial power in the subcontinent. By the 1880s, it was clear that agendas for controlling the spread of disease were akin to colonial interventions into civic life.[3] IMS policy sought to define and delimit those aspects of Indian life penetrated by medical apparatus, and it became clear that where the state was active, indigenous medical practitioners were marginalized and pushed out. Native practitioners had a medical role within the state only as authorized agents in the new bureaucracy, collecting and collating data, or administering treatments in accordance with precise IMS instructions. It's in this instance, however, that the limits of the reach of colonial state are made somewhat more evident: while the surveillance of the population had a hegemonic effect on the conceptualization of the Indian body, the maintenance of the body in the face of common, non-threatening illness and disease remained rooted in the medical practices that predated the introduction of biomedicine. These practices, however, and the structures that ordered them, were not impervious to change, and responded in complex ways and with varying degrees to the development of a public health agenda. Taken together, both the biomedical and biomoral systems of medicine underwent contingent (though often contradictory) processes of change in the late nineteenth century, therein complicating our notion of 'colonial medicine'. Medical historians have predominantly attempted to limit discussions of colonial medicine to the actions of the IMS and other GOI bodies. However, a survey of the indigenous medical systems during this period reveals the ways in which medical systems working outside of this framework had to be reinvented in the context of colonial rule.[4]

As discussed earlier, the figure of the native doctor was a subject of great contention for both the Orientalists and the Company in the early half of the nineteenth century. The debates about the medium of instruction, the content of curriculum, the adoption of certain techniques and the efficacy of indigenous treatments together constituted a nuanced discussion concerning the training of indigenous practitioners.

The notion of the 'native doctor' was itself under frequent contestation, taking on different meanings for different factions of the colonial government. While the term native doctor originally referred to indigenous medical practitioners, after 1826 it referred to those enrolled at the Native Medical Institution, and after 1833, it was restricted for those who had passed the exams at the Calcutta Medical College. The growing rigidification of the term and the power vested in those who could lay claim to it characterized the larger health strategy of the Company, namely the dissemination of European medicine to the people of India through key Indian and Anglo-Indian practitioners trained in it. However, the reorganization of the public health agenda in the 1850s put the eradication of epidemic illness at the centre and demanded that medical expertise conform to the measures necessary to eradicate potential medical threats to the stability of empire.

The ramifications of these changes for practitioners of the indigenous medical systems, represented at one extreme by this legislation, included their failure to enjoy the special relationship newly created between the state and the practitioners best suited to carry out its revamped policy. Marginalized from the ranks of the IMS, vaids and hakims held no place within the scope of state medicine and were forbidden from encroaching upon the 'domain' of the IMS. However, in some instances, colonial officials needed to rely on indigenous practitioners for access to Indian medical knowledge. In these situations, they attempted to cast vaids and hakims as individuals with access to society but without the power of a sophisticated medical infrastructure. This was particularly true during the vaccination reforms of the 1880s, where native practitioners were recruited as surplus vaccinators.

Despite their enforced absence from the discourse of state expertise, the position of vaids and hakims was rather unique. Other indigenous practitioners with less traditional legitimacy could more easily be incorporated into the colonial state. In particular, *dais* (midwives) were mobilized by colonial public health schemes that sought to identify and reform their skills. An extensive historiography has developed around the figure of the midwife in Indian colonial society, tracking her demonization in the mid-nineteenth century followed by her position at the heart of social and medical reform debates and efforts, particularly through the efforts of Lady Dufferin and the foundations she endowed.[5] This narrative makes evident the ways in which *dais* were professionalized through the education and regulation of their practice by the colonial state, and in so doing were stripped of their cultural and social power in traditional society. Interestingly, the introduction of these attempts at reforming the *dais* through education created a space for public discourse to emerge. For instance, the *Garbhrahasya (Protection of the Womb)* of a Dr Chakraborty was ostensibly a guide to conception, gestation and birthing, recorded by the author for use in the new institutions set up to professionalize *dais*.[6] The deviance of the instructions from the technical terms for the body and the reliance upon domestic practices related to food preparation, amateur medical advice calls into question the validity of the text as an authoritative biomedical guide. It leaves the reader pondering as to the in-depth knowledge the author seemed to have of the female-dominated practise involved in producing middle-class Bengali domesticity. Did Chakraborty teach the *dais*, or did he merely codify their knowledge for their reconsumption through the framework of their colonial education?

At the same time, the decline of Mughal cultural authority in the wake of political collapse meant the virtual erasure of the last bastions of Mughal patronage, especially in former imperial centres like Lucknow and Delhi. One approach, adopted both at Aligarh and in the intellectual circles of the young Delhi elite, was to incorporate Mughal knowledge with Western learning, a move towards hybridity that employed a teleological view of the potential value of imperial knowledge in the intellectual spheres of India.[7] However, these organizations with their routes in imperial science had little to offer hakims. However, certain hakims made use of their former courtly ties (or those of their families) to reinvent themselves in new and interesting ways. For instance, Hakim Ajmal Khan, scion of the former Hakims of the Delhi Court, moved in 1892 to the princely state of Rampur, where he was employed by the Nawab as his private practitioner. In Rampur, he was given the task of ordering the Nawab's vast library, which included many rare Islamic medical manuscripts.[8]

At the same time, and as we saw in Chapter 1, the IMS at times lay at the mercy of indigenous medical practitioners. The aim of the IMS in the nineteenth century was to offset any disease or medicine-based causes for the disruption of imperial governance or economy. The primary concern of the IMS – and the impetus for various early health acts passed by the East India Company (EIC) – was the health of the army, under attack from venereal disease and more generalized occupational hazards. Eventually, health policy translated into the protection of the Indian masses, whose health and living conditions were understood as a political issue. The ill health of the masses was bad for the imperial economy, due to the loss of labour and the cost of healing; at the same time there was substantial fear that the ravages of disease might lead to revolt. The IMS' solution was to inoculate the population preventatively and to treat diseased subjects during times of famine or epidemic.

From the onset of imperial interest and concern with Ayurveda to its eventual dismissal and erasure from imperial life, Ayurveda had come to be constructed in imperial discourse as illiberal, unscientific and ahistorical: writ large, Ayurveda was nobly ancient but irrelevant. During public health campaigns, however, when IMS officials were forced to rely upon Ayurvedic practitioners to implement medical policy on the ground, Ayurveda, through its newly useful practitioners, was assigned a new set of characteristics: trustworthy, familiar, locally relevant and soundly reliable.

### Indigenous drugs committee

In 1899, the Society for the Resuscitation of Indian Literature published a book entitled *Ayurveda; or the Hindu system of medical science*, written by Manmathnatha Datta, and published in Calcutta that year.[9] In it, the author put forth the idea of an ancient, textual tradition that formed the basis for the medical traditions of the Hindus. In it, he cited Thomas Wise's works, and referred to William Jones' work, situating him firmly with a neo-Orientalist position. In essence, the text did what so many others had done before it: it recast Ayurveda as a timeless, ancient tradition that resonated in the worldview of contemporary Hindus and that was also divorced from contemporary medical experience. More than 100 years after William Jones and his colleagues had identified the Indian contribution to the global history of medical development, it was clear that, from an imperial standpoint, the image of Ayurveda would remain inescapably ancient.

The Indigenous Drugs Committee was founded within the context of a scientific endeavour divorced from the practice of medicine and instead concerned with the state of the public health. The proposal to form a committee was brought forward to the GOI in December 1895 following a meeting of public health practitioners at which eight papers promoting the uses of indigenous drugs were read to much support from the membership. Moving forth on the basic premise that 'it be recommended to the consideration of the GOI that an extended use of indigenous drugs is most desirable', the committee made three suggestions vis-à-vis this new plan of enquiry: '(1) that definite pharmacological preparations of certain indigenous drugs should be made at the Medical Store Depots for distribution to the various hospitals and dispensaries for trial and report; (2) that medicinal plant farms should be laid out in the districts most suited to the plants which it is supposed to grow; (3) that a drug emporium for India should be established at Calcutta'.[10]

The major rubric under which these concerns were posed was the emerging field of bacteriology, and its potential use to the GOI as a way of treating imperial public health issues through local means. A Dr Hart (who was then editor of the British Medical Journal) concerned about the spread of disease in society and the expense that the treatment posed wrote to the government. The examples cited included cholera outbreaks in Burmese jails, the spread of typhoid through contaminated water in Mandalay and the contamination of water supply in Madras despite the strong presence of the Indian Medical Services there.[11] The discussion was fundamentally about bacteriology, and whether or not it could be successfully researched in the subcontinent. However, as the decision to invest in it was taken, larger questions about the staging of allopathic strategies using indigenous techniques started to take form and a conversation was initiated about the borderline between biomedical and biomoral systems.

The concerns about contamination and the push for localized bacteriological research raised the level of awareness already extant about the nature of epidemic diseases. Prof. Hankin, a bacteriologist based in Agra, had identified the spread of cholera at pilgrimages and *melas* in Allahabad through contaminated water, research that Hart felt had resonance not only in 'furnish[ing] valuable and definite information as to the places and manner of origin of epidemic outbreaks at great popular gatherings, but in explaining the natural history of the growth and intensity of the cholera virus, have a far wider application throughout the empire, and furnish knowledge of permanent utility to sanitarians throughout the world'.[12] These sorts of scientific endeavours were, as Hart put it, 'so clear and striking evidence of the invaluable application of skilled bacteriological research to the everyday occurrences of Indian sickness and mortality that [he] need not dwell upon it'. Bacteriology was the new key to public health, as vaccination had been nearly 100 years before. Englishmen trained in the great medical institutions of Britain could venture out to the colonial laboratory where 'the virgin soil of Indian bacteriological research would offer special opportunities'.[13]

The proposal of an Indigenous Drugs Committee was meant to push the boundaries of what had come before by both the major intellectual expansion of the endeavour beyond the bounds of earlier imperial pharmacological research and the early discussion of the economic value of an indigenous drugs industry. Indigenous drugs had been a part of tropical medicine from the point of contact between Europe and Asia and had an important role in framing public health policy for the EIC dating back to the eighteenth century. The creation of Indian Pharmacopoeias in the nineteenth century and the addition of Indian flora and fauna to the British Pharmacopoeia had created a systemic model through which indigenous knowledge could be incorporated into British medical and scientific thinking. Ultimately, the idea of an indigenous drugs industry was abandoned, and did not resurface until the early post-war period, when the indigenous drugs industry could be reframed as an economic necessity in the face of post-war global drug shortages.

Shula Marks' question about the 'colonial' aspects of colonial medicine, which was explored in the Introduction, provides an interesting framework for thinking through British interests in pharmacology, as the justifications for the committee make clear.[14] In a letter from Babu Bolye Chunder Sen to J.F.P. McConnell, who later headed the committee, the acquisition of pharmacological knowledge in the past was called into question in rather interesting terms:

We may be pardoned for venturing to say that the books now extant in English, on indigenous drugs, by eminent medical men, are no doubt valuable to contributions to medical literature, but are of little practical value when seriously considered; for they are not the embodiment of the experiences of those gentlemen, but are mere translation of the therapeutic uses of drugs from Sanskrit works by the help of Kobirajes without being tested as to their real value by personal observation. Medicines cannot singly be very efficacious in combating diseases with their various symptoms and complications so a mere translation, in our humble opinion, will fall flat on the profession like their predecessors, and meet with the same fate.[15]

According to Chunder Sen, the 'colonial' framing of medical information relied upon the production of knowledge made popular in the late eighteenth century and ultimately upon native informers who coded knowledge but were unconcerned with its widespread application. Ultimately, while this knowledge was theoretically interesting, Sen pointed out that it would not necessarily stand up to rigid testing. The answer instead lay in the employment of 'native gentlemen' – clearly referencing the 'native doctor' of the 1820s discussions – who were properly well regarded socially, who professionally might recommend drugs known to them to the state for the purposes of testing their efficacy and who might have their names submitted 'for their services in the cause of science' for the labour and effort such activities might require.[16] A Kobiraj Sham Kishor Sen of Babooram Ghose's Lane, Calcutta, was willing to test medicines on his own patients first before preparing compounds for the committee, a model that Chunder Sen thought might be an appropriate one for the recruitment of practitioners.[17]

Sen's model never became the stated norm, but it very much reflected the concerns that the government had with the sort of research that had already been done and the ways in which they wanted their intervention to move the field forward. Questions of dependence on 'local expertise' were raised frequently in conversations held by the Medical Board, and it was suggested that both native and European local experts, as identified by local committees, be consulted. The debated role of the 'native' practitioner resulted in re-evaluation of the scope of the endeavour. The possibility of local committees that would engage with local networks of practitioners served to decentralize the project, and expand it exponentially by province, allowing for a much larger survey than had been initially envisioned. This had always been part of the plan, dating from the second meeting of the committee, where medical officers and surgeons in the Northwestern Provinces and Oudh associated with the medical college at Agra were expected to have 'special opportunities of consulting *Hakims* and *Baids*'.[18] Furthermore, the outcome of the research, if filtered through native practitioners, was also much larger than had been intended, as the recommendations it made could be passed back down to the village practitioner-cum-informant, and allowed the project to 'be viewed in the light of the teeming millions of population'.[19]

This role of the practitioner was further expanded upon in a rebuttal to Watt, which had suggested that the mere identification of drugs was not enough of a step away. He argued that biomedical students trained in the treatment of disease would always attempt to procure the sorts of medicines they had been trained to use, and wouldn't switch easily to new drugs. The challenge was, therefore, to replicate compounds that would provide a substitution for a foreign drug – and not an alternative to it.[20] The economic benefits were enormous – C.J.H. Warden reckoned that twice as many dispensaries could be supplied for the same cost.[21] His plan was to have native students studying at medical colleges test out the indigenous drugs in their college laboratories, using Indian products to the exclusion of all European vegetable products.

Warden's plan is of interest as it points to the inability of 'natives' to move away from their racialized status, regardless of their allegiance to the imperial state, or their training in allopathic biomedicine. When George Watt warned that 'the drying and collecting of indigenous drugs is not properly understood by Natives of India and adulteration is largely resorted to', therein discrediting Indian agency in pushing forward the field, he was not, ostensibly, referring to those Indians admitted to and trained at GOI-funded and approved medical institutions.[22] However, even those Indian students who were admitted to the hallowed halls of a colonial medical institution were re-inscribed as natives when it came to the academic mediation of their literal environment, made evident in this example, in the flora and fauna they were charged with dissecting and observing in the laboratory. Regardless, the responsible 'native' in question was a loyal subject of government working through its institutions, but whose ethnicity could be easily deployed by a state wanting to tame the natural world until it conformed to the logic of government.

The other 'native' in question was the one who appeared in discussions of collecting and identifying species of plants and varieties of cures extant in the 'indigenous' practice of medicine. Throughout the collection of compounds thought to be useful to the development of a drugs industry, and that were subsequently tested in the medical colleges of the colonial provinces, throwaway comments were made regarding the individuals who found them or used them already. The committee's reports consistently denigrated their native informants, accusing them of mishandling the materials found and used, or of not understanding their more valuable uses. This cast the traditional practitioner as a figure who had merely stumbled upon a great find, but who was unequipped to make valuable use of it.

The Indigenous Drugs Committee had little concern for the practice of indigenous medicine. Implicit in the implementation of its mission was the divorcing of indigenous knowledge from the context in which it was produced, and its subsequent reframing as Western science. However, the committee spent ample time negotiating the role of native practitioners in order to identify the medical interventions it could make. Regardless of the continued representation of native informant as being rather accidentally imbued with a sense of scientific awareness, the committee's focus on the Indian mediators – be they accidental collectors or subservient laboratory technicians – makes evident that Chunder Sen's admonishment was taken seriously. The focus of government discourse had shifted from the textual tradition to the experiences of native practitioners. Gone were the days of relying upon pandits, translators and ancient texts to interpret medical knowledge. Ushered in their stead was the era of engagement with the contribution that those individuals well versed in daily health practices had to offer the state.

## Practitioners as agents of change

The applied biomedical focus on the practice of medicine inspired an enquiry into the experiential and evidentiary support for new investigations, and native practitioners held the key to this sort of information. The Indigenous Drugs Committee's negotiation of the role of the practitioner marked a break away from earlier approaches to dealing with the Indigenous Medical Systems by privileging the experience of the practitioner over the authority of the text. As we saw earlier, these two spheres had been rather separated from each other, as those who were familiar with the texts were often not involved in the practice of medicine and vice versa. The GOI, however, was not the only body making this transition – in fact, we see this reworking of the organization of Ayurveda by practitioners themselves, especially in their interactions with the state.

The Akhil Bharatvarshiya Ayurved Mahasammelan (ABAM, All-India Ayurvedic Congress) was formed in 1907 by Shriyut Pandit Shankardaji Shastri Pade of Allahabad and consisted of a membership of pandits and their supporters throughout the subcontinent.[23] Ostensibly, these were the Brahmanic keepers of tradition that were being left out of ventures like that of the Indigenous Drugs Committee and were no longer being consulted on matters concerning the modernization of traditional knowledge. In an age where the collection and dissemination of knowledge through the elite was quickly becoming redundant, this attempt to collectivize was clearly about goals other than social service. At the same time the impetus for collective identity-building, especially amongst vaids, was at once linked to a very different series of processes that had little to do with the colonial state. As Carey Watt and others have made evident, a popular articulation of nationalist consciousness in the early twentieth century, in part inspired by the *swadeshi* movement, was a commitment to the ideals of service (*seva*), uplift and community-building through the development of social and charitable associations.[24] This new focus on associationalism was a part of the development of a national consciousness in a period during which the Indian National Congress was particularly preoccupied with upper-level political dealings, as Watt has made evident.[25] The terms of collectivization around indigenous medicine, however, were somewhat different. The ABAM's *raison d'etre* was not the uplift of society, or the practice of proto-citizenship, as it was for social groups. Rather, it was the coming together of a particular set of individuals, linked loosely by their participation in the production of a very elite variety of knowledge, to claim some control over the way in which their livelihood was being reformed and renegotiated in this period of change.

The primary goal of the ABAM, stated clearly in its Memorandum of Association, was 'to develop a form of Ayurvedic polity to secure an effective hand in the control of the State Medical Department to enable the formation of Medical and Health Departments in the provinces on the Ayurvedic lines which would both bring about a general recognition of practitioners of the science and give them an opportunity to take steps or evolve schemes or pass legislatures suitable to their purpose or needs'.[26] Their other goals included the 'fostering of mutual friendly relations' between practitioners throughout the subcontinent, and between Ayurvedic institutions, and the centralization and regulation of practice throughout the subcontinent. The Mahasammelan clearly served as a professional association, set up, as stated in their mandate, 'to induce government, local bodies and other authorities to grant such rights, privileges and concessions as may be necessary for furtherance of the objects of the AIAC [ABAM] and the general welfare of the people'.[27]

The ABAM's focus on textual accuracy and translations of archaic *shlokas* reveals how out of touch the organization was with the new norms of civic identity. For instance, throughout the 1910s, the group had been involved in the creation of an anatomy textbook, called the *Pratyaksha Shariram*. It was meant to advance the reform of Ayurvedic education, a stated goal of the group, and the pandits hoped that it could be the basis for the curriculum of the All India Ayurvedic College it strove to build.[28] This makes evident that the members were conscious of the changing times, but they failed to intervene meaningfully. The idea of a textbook put them squarely in line with some of the other changes that characterized the modernization of Ayurveda. However, the members were not all pleased by the translations done by Gananath Sen, the main author of the text, and the issue apparently split the ABAM.[29] Another version of the book, authored by Pandit G. Shastri, one of the main critics of the original text, had been published in the early 1920s by way of rebuttal. During a period of illness, Gananath Sen relocated to Darjeeling, where he wrote a pamphlet engaging the points of criticism brought up by Shastri and offering new insights into the ancient works of Susruta.[30] By 1930, the matter still had not been resolved, and the ABAM meeting in Karachi still revolved largely around this matter.[31] The ABAM clearly deployed the trappings of colonial modernity through its performance of associationalism, through annual meetings and through the development of a series of goals and dictums for the group. However, its concerns failed to grasp the attention of government, and it was routinely ignored, both during the early decades of the twentieth century and much later as well. Its focus on education and the 'welfare of the people' was in line with the social development missive common to other contemporary organizations, but its plan of action failed to resonate with the voluntary spirit of the times. The Akhil Bharatvarshiya Ayurveda Mahasammelan got the model right, but quibbles over issues of language diverted attention away from issues of social development, and the text was never circulated outside of the context of the group. The group ostensibly missed the altruistic basis upon which other organizations were able to leverage for power.

However, not all indigenous medical organizations were as naïve or misguided as the Mahasammelan. Though some Ayurvedic pandits had failed to engage meaningfully with the new social order, other practitioners, such as a group of Unani practitioners in Delhi, were more successful. Backed by already developed networks and a sophisticated set of institutions, these hakims were able to galvanize the infrastructure already in existence, deploy the language of biomedical progress and represent themselves as the arbiters of the Indigenous Systems of Medicine (devoid of community, ethnic or religious affiliation). In so doing, they were able to take advantage of new opportunities for advancement.

The most notable of these was Hakim Ajmal Khan, founder of the Unani Tibbia and Ayurvedic College in Delhi. His identity as the descendant of one of Delhi's oldest and most important courtly families connected him to government affairs as a representative of traditional networks of indigenous authority, though Ajmal Khan was keen to act as a strategist concerned with the public health of a modern nation. The family had formed an integral part of the Mughal Court and had opened a private practice and institution for the study of Tibbia medicine in Delhi in the early nineteenth century. It rapidly attained a position as a culturally cohesive and socially regulatory body of Unani medical practice: for instance, though Unani generally remained beyond the attentions of government, the Tibbia Madrassah was a significant enough structure so as to merit a dedication by R. Clarke, deputy commissioner of Delhi, on 23 July 1889.[32]

The Tibbia Madrassah occupied an important cultural and political space in the changing politics of the late nineteenth century. Despite its courtly associations, its founders had criticized the inaccessibility of court medicine for common folk and had opened its doors to the urban poor from the beginning of the nineteenth century.[33] Further-more, in an era of public health reform and the imperial denouncing of Indian-run health services, it incorporated allopathic modes of medical learning, often sending its students to government- and privately run schools of allopathic medicine in order that they be educated as widely as possible. At the same time, Ajmal Khan insisted on the inclusion of indigenous systems besides Unani medicine and incorporated Ayurveda into the curriculum of the college, hiring pandits to teach ancient Sanskrit texts. He also went so far as to rename his institution the Unani Tibbia and Ayurvedic College of New Delhi.

In 1912, Ajmal Khan put in a request for a grant of free land upon which to build this Tibbia College.[34] Its popularity and perceived importance inspired its governing body to move and expand the school. He did it under the rubric of a group he had put together, called the Anjuman-i-Tibbia, which represented the Tibbia Madrassah, the Madrassah Zenana and the Hindi Dawakhana (medicine dispensary), thus combining three centres of medical authority into a unified infrastructure. He also assured the GOI that he could do his own private fundraising, naming the Nawabs of Rampur, Tonk and the Begum of Bhopal as his major sponsors, and accounting for Rs 2,31,800 that had been collected to build the college. The Anjuman requested the grant of 100 acres upon which to build a new central site for the college. This was one of the first requests put forth for land to be granted to a charitable organization for free in the New Capital, and the request thus raised two issues for the GOI: firstly, the policy of allotting land in the newest, and most important, city in India, the cost of whose construction had not yet been repaid; secondly, the policy on indigenous medicine, which was in the midst of being reformulated. On the first point, whereas land had been regularly donated to 'public bodies and private individuals for public and other purposes', the selling of land in New Delhi was crucial to the recouping of the costs of both securing the land originally and building upon it.[35] It would not be doled out for free in New Delhi. As the authors of this report went to great lengths to point out, New Delhi differed from other parts of the subcontinent in so far as the land that constituted the city had been bought at great expense, both financial and political, and was therefore different from the majority of land in India that had been held from 'time immemorial' and the cost of which was decided upon using very different indicators of value.

This request for free land required that the Medical Department, which had received the request first, create a formal policy on the difference between charitable and financial endeavours in the newest of

Indian cities and determine at what level the state would be involved in different aspects of public life. What was the meaning of public use and should the state continue to support it financially at the risk of a loss of profit? The provisional answer given within the discussion was that charitable endeavours would now be charged for their land, unless their cause was exceptional. Indigenous medicine was not thought to be a cause that qualified; in fact, as the writers of this proposal pointed out, grants of land for Christian religious purposes would only be made upon the receipt of a fair market price, and it is in all likelihood that indigenous religious traditions would only receive lesser treatment. Moreover, Ayurveda and Unani failed to fit neatly into one specific category and instead were spoken of as religious, educational and nominally scientific entities, making it more difficult to determine their exact use in society, and therefore more difficult to justify the funding of institutions in which they would be practiced. For instance, it was finally decided that the Department of Education should be the final arbiter of the 'usefulness' of the college to public life; however, the size of the land requested was compared to the Bishop of Lahore's request for a Cathedral and accompanying buildings, and ultimately the Department of Public Works was asked to determine whether the request of 50 acres was too extravagant.[36]

Interestingly, the Public Works Department failed to determine whether the request was fair or not, and sent a letter to the Education Department asking for advice. Ultimately, the Education Department felt that taking a decision was beyond their capability as medical schools were dealt with by the Home Department's medical division. At the same time, the wealth of the charity that initiated the proposal, the Anjuman-i-Tibbia, was assumed to be considerable, which in turn undermined the appeal by dismissing its excessive land request as part of the committee's plot to construct 'palatial' buildings for which they must be financially liable.[37] Moreover, it was thought that the sprawling style of land use apparently favoured by the Anjuman-i-Tibbia defied the 'styles worthy of the New Capital, which this association is in the least degree likely to be able to undertake or efficiently maintain'.[38] Thus from the beginning of its political career in the late colonial period, indigenous medicine shifted between religious, cultural, scientific and educational categories, defying the logic of colonial organization.

The government was forced during this process to recognize the social significance of the Anjuman-i-Tibbia and the medical systems that it promoted. While the government eventually decided against granting land, it did give the question serious consideration. After nearly a century of dismissing indigenous medicine as backward and unscientific, and after several attempts to close down institutions and disallow practitioners to practice in any significant way, the imperial government was moved in the 1910s to rationalize Ayurveda and Unani as effective medical systems worth funding. The committee argued:

> The indigenous medical school at Delhi dates from Mughal times and is held in considerable esteem by the Indians in different parts of the Indian Empire. Lately the tendency of the school has been to admit some of the accepted doctrines of modern medical science as legitimate subjects of study for the professors and pupils of the ancient Ayurvedic and Unani systems and the English systems of surgery and bacteriology is definitely included in their Educational curriculum... Without pretending to discuss the merits of the Ayurvedic or Unani medical systems, they believe that the regularization of instructions in these systems, especially in view of the readiness of the Anjuman to recognize the necessity of including surgery and bacteriology in their curriculum, cannot but have a positive value.[39]

The dichotomy of value marks a departure for the GOI in a significant way. Nonetheless, the position of the government also makes evident the extent to which Indian organizations, especially those that ran parallel to pre-existing British institutions, were required to adopt a British worldview about tradition and modernity. The Anjuman-i-Tibbia's request was considered because it worded its request in such a way as to privilege the Western contribution to Indian medical systems, thus restructuring indigenous medicine along allopathic structures of practice and theory. Despite reservations about both indigenous medicine and allotment of land, the Anjuman-i-Tibbia bought 50 acres worth of land for Rs 15,000, and it was agreed that an additional 15 acres would be held for the college in order to allow for further development of the college grounds, when necessary or when financially viable.[40]

As we have seen, the imperial government, backed by the Indian Medical Service, had uniformly dismissed the Indigenous Medical Systems when any discussion of their general or widespread use arose, though it turned a blind eye to the practice of indigenous medicine outside of the official sphere of government. In 1912, the GOI was forced to re-evaluate its position following a request that the Viceroy lay the foundation stone for the Tibbia College in Delhi and accept a deputation of indigenous medical practitioners who wanted to discuss the status of indigenous medicine in India.[41] Ajmal Khan noted that most Indians, regardless of social position, were familiar with at least one of the systems of indigenous medicine and trusted them and that they were based on sound principles of science. Regardless, Ajmal Khan's and his Anjuman's desire to make indigenous medicine palatable to those deployed to spread the Western medical tradition in India was firmly rejected at the higher levels of government: in the second of several letters of this sort, C.P. Lukis, the director of the Indian Medical Service from 1911 to 1917, strongly advised the government to ignore the requests of the Conference. His response is particularly interesting in light of the letter he received from the Viceroy's office requesting his advice, which went to great lengths to cite the stance of both the previous directory of the IMS as well as C.P. Lukis' position in 1910, and specifically asked Dr Lukis if he cared to reconsider, especially in light of the new Medical Registration Acts that were being debated and would soon be in place.[42] Dr Lukis, maintaining his earlier position, mentioned that 'Tibbya' (he placed the word within quotation marks) was unempirical, rooted in superstition and of no concern to the organization of medicine and medical institutions in India.

The imperial government raised a number of political objections to the institution, most notably commenting upon the potential for communal unrest. It was decided that, due to potential unrest in Punjab, the Viceroy should neither lay the foundation stone nor receive the committee. The King Edward Memorial College and hospital scheme in Lahore had promised its students some instruction in indigenous medicine. The administration understood this to mean a basic course in the work in which the Indigenous Drugs Committee was engaged, including the classification of botanical species with medicinal value, and the evaluation of the potential of medical conditions treated or identified by indigenous practitioners. The students at the school were disappointed not to be taught the Ayurvedic and Unani systems of medicine, and were left feeling bitter about the experience. The government thought it unwise to upset them further for fear of political unrest of a communal nature. The government also chose to view the college through the prism of official preconceptions about communal tension.

Official correspondence on the question mentioned several times the fear that the predominance of Islamic benefactors and Unani Hakims on the committee would lead to a serious rift between Ajmal Khan and the vaids included in his deputation, and it was felt that without his participation the Tibbia College would not last longer than his period of involvement with it.[43] Ultimately, the Viceroy decided that he would lay the stone, despite the misgivings of his committee. Regardless, he made it clear that he did not believe in the Indigenous Systems of Medicine, though he did acknowledge that there was some good in them, and that the masses relied upon them for basic medical treatment.[44] He refused the deputation, however, on the grounds that there was no scope within government policy and planning for the inclusion of vaids and hakims, as would be made evident in the Medical Registration Acts that were on the docket.[45]

## Constructing authority through regulation

This section will explore the regulatory codes and registration efforts that were adopted in 1910s at both the national and provincial levels, creating lists of 'valid' treatment centres that the imperial government would fund. The process ultimately backfired on the government: in attempting to regulate care by refusing to fund any clinic or practitioner that could not be termed allopathic, it identified huge gaps in its public health plan that had been filled by the employment of non-allopathic treatments and consultants. Significantly, it was the local IMS officials of Poona in 1915 who begged the government to allow an important Ayurvedic dispensary to remain open and to receive funding, as its presence was crucial to public health practices in the region. These regulatory measures provided little information about the actual make-up of health-care practices in the subcontinent and only served to illuminate the limited capacity of the GOI in this area.

The explicit policy of the GOI was to restrict access to the prestige of European medicine, creating regulatory acts that granted some but not all powers – and financial opportunity – to those who were not trained in allopathic medicine. While the mass appeal of Ayurveda and Unani medicine seemed too large a factor for a government preoccupied with public health to discount, strongly worded legislation that clearly delineated the qualifications of official medical practitioners ensured the continued marginalization of indigenous medical practitioners from health schemes throughout the provinces. Across India, during the 1910s, the rhetoric of the imperial government came into conflict with the practicalities of local health-care provision. Paralleling the interplay between government and Delhi local geography was the relationship between officials and medical services. In practice, however, the authority granted to allopathic and indigenous practitioners in local contexts was blurred and power did not operate according to the determinants of national government.

From the early 1910s, a series of Medical Registration Acts were developed in various provinces. A National Act was created in 1915, entitled the Indian Medical (Bogus Degree) Bill, and drafted by C.P. Lukis. The purpose of the bill was to prevent 'the grant to unqualified persons of titles implying qualifications in western medical science, and the assumption and use by such persons of such title'.[46] The act defined biomedical science as 'western methods of allopathic medicine, obstetrics and surgery, but does not include the homoeopathic or Ayurvedic or Unani systems of medicine'.[47] The fine for laying claim to being a Western medical practitioner was set at Rs 500, and it was also stated that only Indian institutions created by an act of Government could grant a valid degree, thus rendering the degrees of many practicing doctors invalid while imposing upon them a very expensive tariff.

The bill was sent out to over 20 prominent Indian and British officials throughout the subcontinent, most of whom supported the bill. Yet there were some reservations surrounding its effectiveness. For example, Babu Brij Nandan Prasad, a prominent lawyer in Moradabad, pointed out that while 'the only way to remedy the evil is to make qualified doctors easily available and to shut out the unqualified ones... (if) the high fees and charges for western medicine continue poor people will have to content themselves with unqualified quacks'.[48] Others had concerns about the status of the institution with which they were involved, most notably the Agra Medical School and the varying level of degrees it granted. Raja Sir Muhammad Tasadduq Rasul Khan suggested that the bill be extended to chemists, while simultaneously worrying about the retroactive effect of the bill on doctors already practicing. Others were concerned with the buying of fake degrees abroad and the proliferation of European quacks in the subcontinent, and with the lack of regulation for chemists. Interestingly, Mr F. Mackinnon of Gorakhpur noted that native gentlemen that he consulted approved of the bill because of its ability to curb quacks who entered into Zenana, and relied on the uninformed testimony of ignorant women 'who are quite unable to diagnose the symptoms of any illnesses'.[49]

Despite the limits imposed by the Bogus Degree Act and the spirit of exclusion that was made evident in its rhetoric, its implementation grew somewhat more complicated in the various regions of the subcontinent. The attempts to amend the Madras Registration Act of 1914 demonstrate the ways in which local medical culture and institutions were integral to the maintenance of healthy communities, regardless of the regulations imposed by the state on certain practitioners. For instance, whereas the National Act marginalized indigenous medical practitioners by specifically excluding them from the scope of appropriate medicine, the Madras Act, and those of several other provinces, took a different

approach and incorporated a clause in the act that forbade the interference with the study of indigenous medicine. Though this did little
to incorporate the indigenous medical systems into the larger health-
care framework, it implied respect for the traditions and delineated their
role as parallel – if not comparable – systems of significance, if only as
a rhetorical device, in the discussion of medicine and medical institutions. However, the system left much room for improvement due to a
lack of transparency regarding the exact role of indigenous medicine, a
problem that resulted in the amendment of several regional acts in the
1910s.

In 1917, the Madras Registration Act came under scrutiny following
the prolonged prosecution of Dr M. Krishnaswami Ayyar, a member
of the Medical Council who violated the act when he took over the
management of a free Ayurvedic dispensary in 1915. Although he did
no work in the hospital, as he knew nothing of the Ayurvedic or
Unani system, and served only as the director of the institution, he
was investigated by the Medical Council, and eventually charged for
'giving countenance to the *vaidyan* who work there'.[50] The council sus-
pended him and threatened to ban him permanently from practice on
the grounds that his 'presence, countenance, advice or co-operation
knowingly enables an unqualified or unregistered person to attend or
treat any patient'.[51] The difficulty in naming the indigenous medical
systems yet not defining their particular role led to the expulsion of
a popular doctor. It also points to notions of authority given to doctors during the period and demonstrates the tension between authority
vested by the state and that bestowed by the subjects of empire, which
resulted in the employment of a somewhat paradoxical logic: Dr Ayyar
was presumably asked to direct the dispensary because of the authority bestowed upon him by both society and state due to his expertise
in allopathic medicine, despite his lack of familiarity with indigenous
systems; however, it was the fact that his authority as an allopathic practitioner lent credence to the indigenous medical systems that eventually
resulted in his suspension. The Bombay Presidency Act was similarly
tricky. A medical dispensary that began to receive state-funding many
years after its inception was found to be run by a vaid.[52]

Ultimately, the Bombay Presidency Act negated the retroactive claim
inherent in all of the Medical Registration Acts throughout the subcontinent, and instead inserted a claim that no other clinics that relied on
indigenous medical practitioners were to be funded by the state.[53] The
state withdrew its funding, and the dispensary had to close. In a surprising if not uncommon move, the IMS officials in Poona petitioned

the government to reopen the dispensary, despite the lack of allopathic involvement in its administration, claiming that the public health of the community would be in decline if this free-clinic was closed.[54] Furthermore, the IMS officials argued that they were understaffed and overworked, and depended on dispensaries, even if unofficial, to play a part in the dispensation of health materials.[55]

By the late 1910s, the GOI's main area of objection regarding Ayurvedic medicine was in the treatment of health epidemics, such as polio or tuberculosis. This initiated a shift in attitude with regard to the Indigenous Systems of Medicine, albeit a small one. This is made evident in Ajmal Khan's attempt two years prior to his original request to send a deputation to the Viceroy. In 1916, with the backing of Sir Reginald Craddock, a long-time member of the Indigenous Drugs Committee as well as the IMS, he approached the government once more. Craddock acted as a liaison between Ajmal Khan and C.P. Lukis, who was consulted about the wisdom of accepting a deputation. Ajmal Khan, through Craddock, made his case more forcefully this time, citing the problems in Madras and Bombay regarding the Medical Registration Act as making evident the need for a rethinking of policy regarding indigenous medical systems and practitioners. Lukis retained his previous defensive stance, noting the IMS' attempt to reinstate Dr Ayyar in Madras and to fund the Poona dispensary, both of which he regarded as exceptions to the rule, and not part of a larger systemic problem with the registration system.[56] Moreover, Lukis felt that Ajmal Khan's claim that the registration systems impeded the rights of vaids and hakims to practice their medicine was unfounded, and also not the Indian Medical Service's concern; he referred this line of questioning to the Legislative Department, which promptly rejected the complaint, claiming that the category of 'Medical Practitioner' was clearly delineated to exclude anyone who did not hold a diploma, and reinforcing the idea that the intent of the act was to determine the responsibilities of those who held this position, while limiting the position of those who did not. While vaids and hakims were necessarily excluded from this category, the Legislative Department maintained that its sole concern was in clarifying the responsibility of its representatives to the maintenance of health in India, and that the plight of vaids and hakims was not pertinent to this issue.[57]

C.P. Lukis held the meeting despite the Legislative Department's reservations. At the meeting, Ajmal Khan brought up the detrimental effects of the registration acts upon vaids and hakims, noting that the rigidity of the new legislation had caused them to lose some of their clientele.

For instance, they were no longer allowed to issue birth or death certificates in several provinces, which had resulted in a loss of revenue for vaids individually and for their associations as well.[58] Craddock, who acted as a negotiator between Lukis and Ajmal Khan, noted that the GOI would give the practice of indigenous medicine recognition when it was regularized, and suggested that a large national institution act as a standardizing body. Even Craddock had to concede that the government could give little support to the project when Ajmal Khan revealed that only a quarter of the promised sums for the building of the Tibbia College had been collected, and the situation was to be rendered more dire due to the falling income of vaids and hakims.[59] The meeting therefore produced few results: the government position did not change, and Ajmal Khan was granted neither funding for his college nor any assurance that the letter of the law would be reconsidered.

Despite the failure of Ajmal Khan's request for funding, the theoretical questions that it raised remained a concern for the government. In November 1916, mere months after Ajmal Khan had been denied funding, the government wrote to the chief medical commissioner of each province enquiring into the state of indigenous medicine, with a view not only to 'improve and encourage these systems' but also, presumably, to decide whether practitioners or institutions of such standing could be in receipt of government aid.[60] The central Medical Department demanded specific information about the methods by which indigenous medicine was taught, the extent of literature available on the subject, the extent to which the systems were practiced and the classes of men practicing them, and the general popular estimation of these systems.[61] The chief commissioners returned unsurprisingly similar responses: backed by local IMS officers, they encouraged the government to abandon the project, arguing that Ayurveda and Unani fundamentally lacked awareness of surgery and of the circulatory system, that practitioners lacked training and that practice could not be standardized.

The evidence presented seems to tell a different story: each commissioner submitted a list of institutions in their provinces where the indigenous medical systems were taught and detailed lists of texts used in these systems that overlapped with each other. At the same time, additional data revealing the names and residences of qualified vaids and hakims, along with the curricula studied at their colleges, imply a systemic bureaucratization of the systems in this period before formal regulation. The representatives from the different provinces unanimously agreed that although popular consumption far outweighed that of Western medicine and would most likely continue to do so, the prospect of government support for the systems would undermine the authority of allopathic agendas that dictated public health and sanitation.[62]

Although the 1916 inquiry had not been sent to practitioners of any sort, local commissioners, in collecting the information required, had consulted with locally prominent vaids and hakims. The Chief Commissioner of Delhi, Sir William Malcolm Hailey, who went further than his colleagues in encouraging the government to standardize and promote the indigenous medical systems, forwarded the report to Ajmal Khan, who wrote directly to the government requesting access to the reports sent in by the other provinces. Though his request was denied, the government solicited an official statement on behalf of his organization, arguing that if held up to scrutiny of Western science, Ayurveda and Unani would surpass all levels of scientific efficacy.[63] With regard to the question of government funding, the Medical Board argued that because of the widespread use of indigenous medicine by Indian tax-payers, the funding of Ayurvedic and Unani medicine should be allotted for in local budgets, and suggested that no committee that excluded practitioners of these systems would be accepted by the Indian population.

In all of these encounters, the figure of the indigenous medical practitioner, replete with local knowledge and authority – and crucial to the continuation of certain allopathic institutions – emerged as the contemporary symbol of an ancient tradition. Arguments about the theoretical structure of the biomoral systems no longer held their water in the face of the active marginalization of indigenous practitioners by state policies. The Medical Department concretized its approach to the Indigenous Systems of Medicine: the practitioner went from being a discursive figure used to underline a more profound point about medical knowledge to being at the centre of debate.

As we shall see in Chapter 3, while attempts at the national registration of practitioners did little to bring indigenous medicine in line with the allopathic practices of the day, the subsequent transfer of medical responsibility to the provinces resulted in the thorough political restructuring of medical practice. The Montagu–Chelmsford reforms of 1919–1921 brought municipal medical activities under closer surveillance by newly formed provincial medical councils that worked to standardize and regulate care throughout the province, further erasing local differences in favour of regional uniformity. At the same time, the expanded cadre of junior officials, most of whom were Indian, entailed a hasty development of officially recognized medical practitioners entitled to provide the various medical certificates, upon which pensions and salaries relied. In the United Provinces, the indigenous medical systems were called upon to fill this need and were rapidly overhauled to meet new standards of uniform practice. After 1923, practitioners of the indigenous medical systems finally entered the age of modern medical bureaucracy.

It was precisely this situation that the Montagu–Chelmsford reforms of 1919 would transform. This change took place within the context of the similar devolution of power to a new Indian political class at the provincial level following the Montagu–Chelmsford reforms, which made provincial governments responsible to a majority of elected Indian representatives. One of the changes made was the transfer of responsibility for the local coordination of medicine to the newly transformed provincial governments, which resulted in the creation of provincial medical councils that would oversee the different municipal activities pertaining to medicine, and would distribute and standardize medical services throughout the region. At the same time, financial responsibility for medical planning, save in the case of epidemics or other crises, would become the responsibility of the provinces.

In 1921 the question of the GOI's support for indigenous medicine was put to the Department of Education, Health and Lands, which served as interim department in charge of supervising the transfer of medical concerns from the Home Department to the provinces. The priorities of the imperial government in the provinces were threefold: to control the costs of health care during a period of its expansion by drawing on pre-existing structures; to secure the cultural legitimacy of Western medicine by creating separate spheres of medicine, and positioning allopathic medicine as that of national importance; preventing epidemics or other health crises by maintaining the dominance of the IMS with regard to public health emergencies or threats. Everyday health matters, however, especially for routine medical problems, were left to the medical council of the provinces, which in the UP was populated mainly by newly elected native representatives, many of whom had recently made the transition from practitioner to politician.

\*   \*   \*   \*

In the first of these discussions after the transfer of responsibilities to the provinces, the Medical Department was asked three questions.[64] To begin with, the Department wanted to determine if it should recommend that provincial governments have Ayurvedic and Tibbia medical colleges established; secondly, it wanted to determine if it should take measures to develop Indian drugs; thirdly, it sought to determine whether hakims and vaids should be appointed in every dispensary to treat patients with indigenous drugs. The directness of these questions makes evident the galvanizing effect of the surveys of the Indigenous Medical Systems a few years earlier. No longer were the claims relegated to matters of legislation – the inquiry in this case encompassed a wide array of social and political responsibilities that, if addressed in a serious way, could potentially revolutionize the health-care system in India. The GOI staved off answering these concerns directly by proclaiming the plight of indigenous medicine to be the concern of the provinces in their adoption of medical planning once the transfer of that power had been finalized. The GOI made clear that despite its shift in thinking about indigenous drugs and aspects of health, it still did not consider Unani and Ayurveda to be scientifically sound, and would not recommend that the provinces place vaids and hakims at the same level as qualified medical practitioners.

This decision makes evident the development of the issue from its inception as a political concern of the GOI to its status in the final stages of its position as a national concern. Whereas in the first decade of the twentieth century indigenous medicine was considered, especially by the IMS, to defy categorization with no serious bearing on health care, by the early 1920s it was the basis for a new industry and was considered an important and viable part of the health-care system. The questions that arose with regard to registration of practitioners, the standardization of practice and the formation of institutions were incorporated into the plans for the provincial administration of health care.

# 3
# Embodying Consumption: Representing Indigeneity in Popular Culture, 1910–1940

In Chapter 2, we saw the ways in which a composite tradition called the indigenous medical systems came to have a more complicated meaning in the eyes of the state, as the GOI encountered – and was forced to contend with – its reliance on the employment of native medical practitioners and indigenous medical practices. The government's Medical Board unravelled the question of the Indigenous Systems of Medicine by dealing with the position of practitioners within them, a strategy that shifted the state's focus on textual authority (as made evident through Orientalist writings on Ayurveda) to the lived practice of the indigenous medical systems. The situating of the practitioner at the centre of tradition – and the reliance on the practitioner as arbiter of the state of the medical system – allowed for the state to interact meaningfully with the indigenous medical systems. This had been impossible when Ayurveda was conceptualized as a tradition accessible only in Sanskrit and guarded by individuals who circulated in spheres virtually uncontrolled by the state. The practitioner ostensibly gave the colonial government access to the current state of the indigenous medical systems, where the Pandit had only been able to provide insight into its theoretical meaning in a broader civilizational context.

A similar process was occurring in the social lives of Indians during the period of the spread of nationalism and the tightening of the boundaries of the nation. Ayurvedic medical knowledge had held an importance in the life of Indians as a part of the broader spectrum of tradition, and the middle class had interacted with it as such, positioning it as a primarily spiritual, cosmic and theoretical rationalization of the way in which the body worked. However, the new social demands for theoretical frameworks through which to understand nationalist ideas created a space for this 'indigenous' knowledge about the body to have a greater social resonance. As individuals who practiced medicine in the vernacular languages, and in contemporary social milieus, these practitioners were already engaged in the social, cultural and political world of twentieth-century North India. At the same time, the broad experience of medical practice represented in these discussions created the possibility for those who 'knew' about the body, even outside of the context of Ayurvedic practice, to participate in them meaningfully. The emergence of a women-centred sphere of medical discussions, which appropriated notions of Ayurveda as part of a larger discussion of indigeneity, makes this evident. The advent of the Hindi public sphere of literary production and dissemination provided the perfect forum for the introduction of a 'modern' Ayurveda into contemporary social discussion by those 'authorities', both professional and not, who could participate in it directly.

It is also in this instance that Ayurveda stopped being a set of contested ideas and started to take on a meaning as a singular system. A plethora of very different authors, practitioners, advertisers and lay commentators published books, pamphlets, journals and advertisements that all claimed to be 'Ayurvedic', which might in a different political context have had the effect of expanding the notion of what Ayurveda was. In the context of growing concerns about national identity, Ayurveda became singularly linked to discussions of indigeneity and the formation of an 'authentic' Indian identity. Thus, Ayurveda ceased to have a multitude of meanings, and a uniform set of ideals came to characterize all of the writing designated as Ayurvedic.

In this chapter, I will look at three specific genres of writing and evaluate the ways in which medical discussions pervaded the field. Firstly, I explore formalized, Ayurvedic writing, aimed at an audience specifically and primarily interested in consuming this information. Secondly, I explore Grhinis, or domestic guides, and other writing aimed at women and dealing with the household. Finally, I explore discussions of medicine, gender and the body in popular advertising in an attempt to expand the parameters of discourse. Ultimately, these discussions reveal the importance of the idea of indigeneity both to discussions of the nation and to discussions of the future. Each field appropriated the term as a key part of its discourse, co-opting earlier usages, but also recreating them for new social, cultural and political contexts. Ayurvedic practitioners and other medical authors employed indigeneity as a temporal concept, laying claim to the 'timelessness' of an 'ancient' Indian tradition, while simultaneously negotiating the appropriateness of Ayurveda for Indian 'modernity'. Those writing domestic guides relied on the notion of the home as the inherent centre of Indian life, casting indigeneity as being viscerally rooted in every-day living practices, daily rituals and local ingredients. Finally, writers in the wider public sphere negotiated these two poles, tempering scholarly opinion with popular imagination, yet retaining at the centre of debate a focus on what exactly it was to be indigenous.

## The origins of publishing and the medical contribution

The advent of a vibrant publishing industry in a newly standardized vernacular language has been identified by scholars like Francesca Orsini, Vasudha Dalmia, C.R. King, Sudipta Kaviraj and Sanjay Joshi as a key facet of the development of a national consciousness, and the self-fashioning of 'modern' identities in the late colonial period.[1] The emergence of standardized, vernacular languages, reflecting the contribution of dialects made to conform to a singular mode of expression, was key to this process. In North India, the lingua franca of the elite literary and commercial classes was called Hindustani/Hindavi, literally the language of Hind, written in Persian script. Persian had been the language used in the context of the Mughal Empire's court culture, but by 1837 even the language of the legal courts had been replaced by vernacular alternatives, making Urdu the official language of state authority. As Orsini and Lelyveld have argued, in the early nineteenth century, people employed a variety of 'linguistic registers' more dependent on social context and status of the speaker than on communitarian affiliations. The government's need to train its officers in Indian vernaculars initiated a process of linguistic standardization and codification in North India.[2] Scholars have pointed to missionary activity and to the development of colonial education institutions, both of actively published language textbooks that relied heavily on the work of Fort William's Oriental linguists.[3] At the same time, new associations between language and communal identity were formed by both colonial administrators at Fort William, most notably John Gilchrist, and activists among the newly emerging elites.[4]

Building on Dalmia's work, both King and Orsini argue that the educational policy promoted by the colonial government produced an increasing number of ambitious intellectuals educated in Hindi in both the Devanagri and Persian scripts. However, their estrangement from the remaining vestiges of Mughal state practice meant that they were not well versed in the traditional language of authority in the region. Frustrated with their exclusion from bureaucratic opportunities of the state that favoured Urdu, these rising intellectuals, mostly of small Zamindar families or peasant elites, lobbied the government of India throughout the late nineteenth century for recognition of Hindi. To promote their cause Hindi advocates increasingly framed Hindi as the 'natural' mother tongue of the people of Hind, and the Devanagri script, and its associations with Sanskrit, as more traditionally and authentically Indian, an effective strategy in the age of Indian proto-nationalism. King has shown that while Urdu in the Persian script remained the dominant literary language throughout the nineteenth century, by the early decades of the twentieth century the tables had turned and Hindi (in the Devanagri script), used in bazaars from Lahore to Calcutta, emerged as the dominant language of oral and print communication by the early twentieth century in North India.[5]

Whereas a vibrant commercial publishing field in Hindi had emerged by the early twentieth century, it was also upon the Bengali traditions of medical writing and dissemination upon which Hindi-language medical writing drew. Several books were translated from Bengali into Hindi, printed in centres in both Calcutta and the metropolises of the United Provinces.[6] Bengali debates about Ayurveda had emerged in the mid-nineteenth-century literary sphere and attempted to reorient the contemporary focus on imperial biomedical systems back to extant traditions in the subcontinent. As Obinosh Chondro wrote in his preface to his *Sushruto Somhito*, 'Do not pine for the alien ship, while your own vessel lies in tatters!'[7] Three modes of publishing have been identified by medical historians of the period: the translation of *Shushruto* and *Choroko Samhitas* from Sanskrit into Bengali, the development of Western-style but Bengali-language pharmacopoeias and 'eclectic' guides to curative, preventive and diagnostic medicine based on local knowledge.[8]

Hindi was in part derivative of its Bengali intellectual and commercial counterpart: Bengali dialects had consolidated into a unified, recognized form of literary expression in the nineteenth century. Bengali had also become a language in which 'modern knowledge' could be transmitted, thus cementing it as a thoroughly modern phenomenon.[9] At the same time, the longer history of Bengali meant that by the late nineteenth century a variegated and sophisticated press, theatre and literary sphere had become entrenched parts of Bengali culture. The many Hindi varieties of regional dialects still at play in the North made the language seem almost backward when compared to the 'advanced' state of Bengali. The break is significant: the mimetic adoption of Bengali 'modernity' by the burgeoning Hindi-speaking intellectual elite reflects a firm rejection of post-Awadhi artistic traditions that dominated the North Indian literary field in the mid-nineteenth century.

## Performing medical literacy

The advent of a literary sphere increased the mobility of the emergent middle classes by providing an outlet in which they could be actively modern, political, intellectual and, most importantly, nationalist in new and innovative ways. This latter category would become and remain the most steadfast characteristic of a literate identity. The nationalist cause would also provide an outlet and goal for the continuation of literary culture. The first two decades of the twentieth century saw the self-conscious fashioning of the Hindi sphere of writing and publishing in the Bengali mode, as well as the drawing of a boundary around who could participate in the public sphere directly.[10] However, by 1920, the publishing industry was very well established, literary societies could look back on decades of participation and the dissemination of knowledge through the Hindi publications (and their subsequent performance for those who could not consume them directly) was an essential part of the production of culture. The growing nationalist consciousness, especially post-war, found a perfect vehicle in the world of the literati and their print culture. The 'nationalist cause' provided a modern challenge very much in line with the project of self-fashioning and identity-building that had solidified the rise to power of certain groups during the period of social change in the late nineteenth century. The development of the genre of historical writing in Hindi dates back to this period, as does the preoccupation with the idea of indigeneity, and the drawing of a firm boundary around the idea of the nation.[11]

Some of the social behaviours associated with membership included the donning of certain sorts of clothing, a rigid adherence to using only the purest forms of language and the insistence upon ideological leaning as the basis for grouping over and above kinship or community ties. These developments relied upon a self-conscious performance of a particular identity embodied in a symbolic break with traditional forms of social interaction. Medical writing, as we shall see later, made use of these developments by deploying similar ideas about nationalism, identity and collective belonging that were being negotiated in these more sophisticated spheres, but couched them in 'scientific' language. For instance, the oral recitation of literary culture in the domestic sphere, or to illiterate audiences, served to increase the scope of the 'public' sphere of Hindi writing. Medical authors seized upon this revamped focus on the home by gearing their writing to address similar themes. However, they not only did seek to participate in the new, modern domesticity, but also sought to exert control over its enactment.

## Nationalism and the body

It is unsurprising that it is in this age of national consciousness medical writing, and discussions that dealt more broadly with the Indian body, emerged as a cornerstone of Hindi publishing. Ayurveda was a perfect portal for these varieties of public discourse. A key trope of nationalist thought, especially in its more informal and public incantations, was an obsession with the idea of precedence, made manifest in discussions of the Indian past that took a teleological view of the nationalist present. This way of thinking venerated all things ancient, lauding systems of knowledge, patterns of social interaction and modes of governance of the *pracheen samay*, the ancient age, as the authentic foundations of the nation. The glory of an 'ancient' age, the *pracheen samay,* was as much a moral location as a temporal marker, bound as it was with the contemporary 'crisis' of colonial modernity. In this, authors envisioned the possibility of a glorious past that could inform the potentially glorious present and future. Of course, their insistence on a golden age was derivative of their forebears, who had tried their hand at social change. As we saw earlier, it had been a common trope throughout the nineteenth century re-fashioning of the nation through efforts at reform.[12]

It is in eclectic medical guides, first in Bengali and later in Hindi, that the really innovative work on medicine and the body began to take shape. Rohan Deb Roy has identified a latent but persistent trope relating to the articulation of a normative and appropriate form 'traditional' and 'modern' couples within guides calling themselves 'Ayurbed' in this period.[13] Debates about maternity and child-bearing, through the lens of the condemnation of the *dai* (midwife), were conducted within books claiming to be guides to indigenous medicine. While translations and pharmacopoeias reified both Sanskritic and biomedical modes of recording medical knowledge, the 'eclectic' guides employed non-standard and individual structures to communicate hybrid, localized knowledge about the body in a new medium. In this context, the body could be used as the site for complex negotiations of myriad topics, beginning with the use of biomoral versus biomedical systems for the cure and prevention of illness and disease, but also about the other discourses in which the body was implicated, like reproduction, sexuality, mobility and travel. In fact, these latter medical guides do not seem to have been based in Ayurveda at all. Though they deployed notions of 'indigeneity' within their pages, they diverged massively from the appropriate form of Ayurvedic writing in Sanskrit. At the same time, authors did not have to be practitioners, and often were not, once again making evident the subjectivity of the author as a new and important trope of modern Ayurveda. Together, the idea of indigeneity, articulated through myriad markers and tropes, triumphed over rigidities implied in the textual tradition of Ayurveda. Ostensibly, the notion of the 'indigenous' eclipsed the 'medical system' in the ordering of this information.

Ayurvedic writing in Hindi began later but underwent a similar pattern of evolution. This was, in part, because writers often claimed the Bengali translations of pharmacopoeias and Sanskrit texts as a part of their own literary traditions, translating these Bengali guides into Hindi in the early twentieth century. At the same time, the more informal guides to medicine and indigeneity were also circulating, both in Bengali and in Hindi translations. Therefore, the emergence of medical discussions in the Hindi literary sphere was, in part, a response to the earlier Bengali experience. The presence of direct translations from the Sanskrit texts, done within the parameters of Brahmanic learning, was few and far between in number, as were straight translations of or compilations of pharmacopoeia. The most prevalent form of writing about indigenous medicine, as it had been in Bengal, was the informal, non-standard guide to medicine authored by individuals claiming expertise about the body, and published by new publishing houses. At the same time, discussions about the indigenous body permeated other genres of writing, most prominently appearing in women's guides to household maintenance and in popular journals also authored by women about their daily lives, making evident the importance of these conversations to the other social and cultural upheavals.

As we shall see, one of the most interesting developments to emerge from these popular medical discussions is the authority given to the author as expert, which points to a reworking of medical authority in this period. Ayurvedic knowledge as embodied in the text had always been the terrain of trained pandits, writing in a particular tradition, and deriving authority from the sort of knowledge they had. The new norms for publishing information about Ayurveda and other forms of indigenous knowledge about the body marked a major move away from this mode of practice. The importance of textuality was retained, as Ayurvedic guides emerged and were consumed in large numbers, and would become, as we shall soon see, a major component of the state control over the indigenous medical systems. However, the articulation of ancient knowledge in a modern language, by those divorced from the practice of authoritative knowledge production, served to put a new expert at the centre. As we shall see, the authority of the author to speak about the body was newly negotiated through his or her ability to say something meaningful about the position of the body in the tumultuous political milieu of the nationalist interwar period in the United Provinces. This furthered the trend of putting the practitioner at the centre of the system and employing him or her as the authoritative arbiter of this traditional system during the period of its modernization.

## Ayurvedic textuality: Practitioner as author

From the 1920s, texts dealing explicitly with Ayurveda constituted an important part of the Hindi public sphere. Earlier Ayurvedic tracts had been written by pandits, in Sanskrit, and had been disseminated as part of the wider process of disseminating religious or ritual knowledge to the public through established social networks. However, the advent of the public sphere created new possibilities for the distribution of knowledge and also provided a way of reaching new publics directly. The earlier reliance on the 'religious' infrastructure that had governed the way in which pandits interacted with the public was now replaced by a norm of communication that didn't rely on this sort of formality involved in these encounters. As the dissemination of medical knowledge and information was more generally vernacularized, vaids now found that they had an advantage that pandits lacked. As such, they were able to reclaim the importance of textuality to the communication of Ayurvedic knowledge, but they were able to reframe it within modern language of social belonging. What we see in the development of this genre of writing is the emergence of the practitioner as the arbiter of tradition, made resonant socially through his reliance on modern Hindi, and made culturally resonant through his reliance on textuality.

Ayurvedic texts were divided into two main genres. The most common was the medical pamphlet, often a short, cheap publication with a large print run that addressed a single topic, or a particular writer's own perspective on Ayurveda. The second variety was the bound Ayurvedic textbook, often offering a conglomeration of different sorts of knowledge and opinion. Though these books seemed to profess very different agendas and address different audiences, in fact their impact seems to have been quite similar, for both laid claim to an informed scholarship targeted at a broad audience. While they deployed different sorts of medical information in varying formats, they were consumed by similar audiences: educated, literate people with an interest in the workings of the body and an interest in nationalism. Together, medical pamphlets and textbooks constituted the formal field of Ayurvedic literature, and drew upon the scholarly traditions being reinvented in the political and cultural spheres of the 1920s–1940s.

The quantity of *Susruta* and *Caraka Samhitas* in Sanskrit remained steady, but they were soon surpassed by over 100 different guides to *chikitsa* (illness), *arogya* health and Ayurveda, all of which were written in Hindi for an audience unfamiliar with Sanskrit. The Hindi employed, however, was of a relatively high standard so as to assume an audience that was fairly literate. At the same time, the structure of the Sanskrit text was retained in the new vernacular: often times, the authors, especially if they were pandits but even if they were not, would print a Sanskrit *sloka* followed by both a translation and explanation of it in Hindi immediately below.[14] This had the effect of rooting claims to knowledge within a learned framework, where authors could lay claim to the authority of the Sanskrit texts through the deployment of it as form, while repositioning meaning as something communicated through the colloquial vernacular.[15]

The other major linguistic trope was the replacement of certain Urdu or Persian words with Sanskrit-derived terms. This process had the effect of reifying Hindi as the language of the Hindus, a process that began in the late nineteenth century and was perhaps made most popular by the Hindi movement's slogan 'Hindi, Hindu, Hindustan', turning a literary venture into a nationalistic one.[16] However, the negotiation of this linguistic space was problematic: often the new terminology, derived from outdated Sanskritic texts or Hindi ritual terms, had no cultural resonance and authors were often forced to place the older colloquial – or even the English term – into a parenthesis next to the new Hindi term. For instance, the authors of a famous Ayurvedic dictionary included phonetic translations for the English (or other foreign language) words that had no Hindi equivalent. For instance, a 1915 guide added a heavily Sanskritized word for fever onto the end of a part of the body to denote infection, and a guide in the 1930s casually deployed Hindi terms for medical fields and conditions.[17]

The 'purification' of Hindi, therefore, was about negotiating the languages of both the cultural and political past and present, evoking for the reader a sense of historical situatedness represented by the form and language of the text. For instance, the upholding of Sanskrit coupled with the rejection of Urdu and Persian terms clearly demarcated an 'appropriate' Hindu past. At the same time, the negotiation of English and Hindi on the pages makes evident the syncretism of categories of 'indigenous' and 'Western' medicine. Yet the 'indigenization' of the language was itself a problematic endeavour: firstly, in some cases, the term could be traced back to the Vedas; furthermore, the reliance on Urdu and Persian language to explain them made evident the entrenchment of these terms within North Indian society, and the tenuous nature of any project set on displacing it.

At the same time, some authors did hark back to 'tradition' by relying on myths that had come to define the ways in which Ayurveda had evolved as a supernatural force. Sections on the life of Dhanwantari and Caraka were included in guides, lending a mythical, ancient characteristic to some of the thoroughly modern material. This was a reference back to a standard practice within traditional Ayurvedic manuscripts, which would trace the genealogy of Susruta and Caraka, and explain how they had acquired their knowledge from the gods. Several authors maintained this practice, but the stories were often cut to a short mention. One author, however, in his *Caraka Samhita* developed an interesting new way of retaining the Sanskritic authority of the guide, while simultaneously making it accessible for non-Sanskrit readers.[18] Pandit Shiv Sharma, a noted Ayurvedacharya who presided over the essay-reading competition at the All India Ayurvedic Congress' meeting in Mysore, used a compilation of printed images to spell-out a mythical story of Susruta's development as a sage, and that of other Ayurvedic sages.[19] Each page contained no more than three images, and often the page contained just one image. The images were drawn by hand and then reproduced, and were fashioned in the style of representation common in popular renderings of gods and goddesses. However, the most interesting thing about the representation of the life of Susruta through images was the almost complete absence of text. Some illustrations included a tiny line identifying the figures in it, but most did not.

### The mechanics of public circulation

The engagement with the linguistic abilities and mores of the readership makes evident the extent to which the writers of these guides were concerned with the cultural and social projects with which their audience was engaged. The focus on the resonance of terms as cultural ideal and their semiotic placement resonated with other literary projects aimed at the middle-class Hindu bourgeoisie. The categorization of disease, the cost of illness, the division of labour around the detection of illness, the application of medicine, the gendering of medical knowledge and responsibility – all of these conform to the norms of reading the social in the inter-war period in the United Provinces.

Journals and newspapers offer more insight, as the temporary and disposable nature, paralleled by the ongoing purchase of this genre of literature, gives insight into the ebb and flow of their consumption. Other authors have gleaned insight into the popularity of the text in this context by evaluating the print runs and price. It is here where the different genres of Ayurvedic literature become important: the pamphlet and the bound book were priced, marketed and printed differently. However, their content and their form remained similarly constructed, despite the difference in intellectual approach. Pamphlets and books seem to have constituted different parts of the same field, albeit with different meaning. Pamphlets, for the most part, were kept at under 100 pages and generally printed on paper. The covers of the text were generally printed on a hard cardboard, and carried little by way of embellishment, besides fancy calligraphic lettering. Print runs seem to have consistently started at 1000 copies for the first print run, often escalating by 1000–2000 depending on the popularity of the book. The price of the volume was usually between 1 and 2 *annas*, making it accessible to the middle-class consumer.[20] The price would often rise by an *anna* with the increased print run, a characteristic uncommon to other forms of literary publishing. The pamphlet can thus be argued to have an economic life removed from that of other literary commodities, like books or one-off publications, whose costs did not seem to change.[21]

Within these literatures, ephemeral literature, produced by authors sensitive to market demand, was most likely to actively reflect the historical moment in which they were produced. Pandit Shalagramaji Shastri's pamphlet on the state of the indigenous medical systems, in which he reproduced a report aimed at the government, arguing for an increase in government spending and recognition of Ayurveda in particular, is a good example of the escalating print run of a popular book. The pamphlet was first printed in 1925 and sold for Re 1 at a time when these issues were just coming to the forefront of debate at the state level. However, a 1931 edition saw the price rise to Rs 2 and the print run to 3000. At the same time, the report, which had been at the front of the book and barely introduced in 1925, was now merely an essay amongst others, most of which dealt with the nature of Hindu medicine. The name of the pamphlet, however, did not change, nor did the billing for the book – though the advertisement inside its covers was slightly different. However, the nature of the pamphlet changed dramatically: whereas the aim of the pamphlet in 1925 was to increase awareness amongst the population about the state of the Indigenous Medical Systems with Ayurveda used as a lens, by 1931 the purpose of the pamphlet was to sway the audience round to the opinion that Ayurveda was both the most appropriate of the medical systems and the one most deserving of the government's patronage.[22]

Ayurvedic bound books occupied a different space within the Hindi Public Sphere. Their print runs were similar to those of the pamphlets, generally beginning at 1000 for the first printing, and increased up to 5000 for some of the most popular books. From the 1920s on, books often included coloured pictures depicting the part of the body under discussion, thus marking a vast difference from the basic print model of the pamphlet. Books were often attached to institutions, like the Banaras Hindu University (which had its own printing press and distribution centre, as well as a centre for Ayurvedic learning). Of all the texts published during this time, Haridas Vaid's *Chikitsa Chandrodaya* seems to have been amongst the most detailed and the most popular. Vaid published his eventually eight-part long encyclopaedia in separate versions under the auspices of his own *Haridas end kampani*, or Haridas & Co. The volume being published rose dramatically from a mere 1000 copies of the 1920 version to 3000 copies per volume in 1930. The price, however, remained at Rs 3 per volume, no surprise as the price per entire collection rose from Rs 3 in 1920 to Rs 24 in 1930, when all of the additional volumes are taken into account; Vaid took this into account, fixing the charge per volume at 4 *annas*. There is evidence of this being an ongoing project in the form of single, additional volumes published individually throughout the 1920s, adding on to the original 1920 edition, though it seems as if writing stopped in 1930 at the conclusion of the eighth volume.[23]

Vaid's book differed from others of the period in both form and content, and it was undoubtedly these differentiating features that secured its popularity. Where other medical books published at this point contained pictorial diagrams, they were normally limited to the part of the body being discussed, and often seemed to have been drawn by hand. Vaid took a huge leap by often reproducing the entire figure of the male body even when only one organ was being discussed, as well as inserting photographs and painted portraits of exhibited medical conditions. These depictions of illness and bodies raise the question of representation involved in a visual rendering: while the rhetoric of medical writers like Vaid was bound up in the racial politics of nationalism, which often served to classify both disease and treatment as decisively Indian or non-Indian, the images used often depicted white men, or white body parts. While the printed and drawn diagrams were carefully demarcated as 'Indian' – made visually evident by a tikka mark on the forehead, a certain rendering of facial features or the inclusion of a 'Gandhi cap' on the head of the figure – there was no accounting for the discrepancy between the rhetorical and the visual.[24]

Another major literary endeavour took the form of the *Ayurvediya Kosha*, the Ayurvedic Dictionary, published by Ramjit and Daljit Sinha of Baralokpur-Etawah from 1938 to 1940.[25] The first volume began with the first letter of the alphabet, *a*, but only made it though the first five letters following *a*, covering both English and Hindi words that began with an 'a' sound. The second volume followed through the alphabet to the first consonant, *k*, cutting down on the number of words per letter, which was achieved by cutting the number of English words explained, though often using Latin or English words to explain a Hindi term. The brothers intended their collection to be a definitive Ayurvedic interpretation of pathology (*rog-vigyan*), chemistry (*rasayan-vigyan*), physics (*bhotikvigyan*), microbiology (*kadin-vigyan*) as well as to the study of deformity.[26] They claimed its uniqueness as the most comprehensive guide to the topic, stating that 'all of the topics, despite the different languages used to describe them, have the same base meaning and play the same role in the conversation about health'.[27] To promote the smooth transition between language and tradition, the authors included phonetic translations for the English (or other foreign language) words that had no Hindi equivalent.

## Ayurvedic notions of national constitution

The agency of vaids in constructing a discourse of this variety is clear. However, the popularity of this genre of writing poses more of a problem. What was it about these texts that resonated? It is here that the significance of Ayurvedic writing as a social project – and not just a professionalizing one – comes to the fore: Ayurveda was recast by medical authors as an inarguably 'scientific' justification for social belonging; in so doing, it was also a recipe for understanding the biological underpinnings of social difference. Rooting social and cultural identity within a newly legitimate and objective 'science', Ayurveda could be employed as an ancient cure for the ills of Indian 'modernity'. In that sense, Ayurveda was not simply a scientific tradition; it was also a reflection of Indian identity and a mechanism for social reconstruction.

Postcolonial interventions in the study of colonial sciences and scientists have identified, as Gyan Prakash has put it, 'the functioning of the language of reason as an idiom of power'.[28] Indeed, the Ayurvedic authors and practitioners used the rhetoric of medicine, science and progress to make authoritative claims about the body; while the points of reference used were locally derived, the structure of debate relied upon the social categorizations of difference. What constituted an Indian body? How was its difference marked from other bodies? How did medicine accommodate this difference? These discussions mirrored the larger concerns about nation, identity and belonging within the Hindu middle class.

The bodies under discussion took many forms, each reflecting the values of the writer and his audience. Discussions of European bodies, for instance, drew on older colonial discussions of race, climate and constitution, adopting arguments about the importance of agriculture to the development of medicine, and linking cure to environment. For example, according to Shyamsundar Sharma, author of a guide to Ayurveda in 1925, 'Even in England, the question of climate is often raised. Between England and France there is distance of only a few miles; but on several occasions, the doctors of France have discarded English medicines, saying that the climate of that place does not suit the patient in the country. And so have the doctors of England rejected the French medicines on the same grounds.'[29] This claim to an essential composition treatable only by the medicine available locally was a common theme amongst those writing medical books for popular consumption in the early twentieth century. However, what makes Sharma most interesting is his invocation of a geographical clash between two Western nations. In refusing to limit the discussion to the confines of the colonial relationship that saw East and West posited against each other in a battle over scientific knowledge, he instead inserted India as a state amongst many, a civilization with one unique constitution, equal to Britain or France or China.

However, the body most often under attack was the Muslim body. This discussion emerged within the context of the communal politics of interwar United Provinces, where some of the most contested battles between Muslims and Hindus were played out. Reflecting these tensions, several Ayurvedic practitioners invoked the medicalization of race to legitimize this viewpoint as scientific, using religious difference as a marker of biology, and these approaches acquired legitimacy as representative of mainstream belief. Pandit Shaligram Shastri, for example, was a very prominent pandit and vaid, chosen by Ayurvedic practitioners throughout UP to write a report for government officials, who were considering expanding funding for indigenous medical schools and practitioners. The report was published in book form in 1931 and was printed at above-average print runs; it contained the text of his report in Hindi, as well as letters of support written by local politicians in English for the convenience of imperial administrators. In the Hindi foreword to the report, he dismissed Unani medicine as foreign and harmful. He claimed that 'while Unani medicine provided particularly useful cures for Muslims, when practiced on Hindu bodies, especially poor ones, it was harmful'. Sharma identified a crucial focus on haemorrhaging or bloodletting in Unani medicine that he felt was inappropriate for Hindu bodies, though was appropriate for Arabian and Persian bodies, and hence, in his reasoning, Islamic bodies. Furthermore, having read Unani texts (though he never names any), he felt that they were, for the most part, inaccurate translations of Ancient Vedic material, garbled by their Muslim translators and peppered with foreign medical knowledge.

For Shastri, Unani medicine and practitioners were fundamentally foreign and biologically different. Their medicine reflected the needs of their constitutions, which he attempted to prove were different from those of Hindu, and thus properly indigenous, bodies. Furthermore, the medical systems that governed their bodies were fundamentally different. His description of Unani blood haemorrhaging rendered the system in an almost macabre light, and subtly suggested that Unani practitioners who might practice this system upon Hindu bodies were sinister quacks. Shastri's perspective represented the exception rather than the rule in this sort of writing; however, the huge print run of even the second and third volume of this text – and his appointment to the Board of Indian Medicine and to a prominent Ayurvedic college the following year – suggests that he was obviously writing something that resonated.[30]

The importance of the critique of the Islamic body was the delineation of a clear distinction between Muslim and Hindu worlds that framed other discussions of difference. For instance, most practitioners conceptualized the appropriate National Body as Hindu – but also as middle class and twice born. The potency of Ayurvedic medicine was considered capable of curing Hindu society of its ills. One of the dimensions of Ayurvedic thought in the late nineteenth and early twentieth centuries was a serious engagement with reproduction. While the idea of conception was mentioned in the *Susruta* and *Caraka Samhitas*, and discussions of menstruation are found in other tracts, a more complicated understanding did not exist in general guides to Ayurveda until the twentieth century. This new focus on the reproductive body and its potential was due in part to the idea of a standardizing view of the physical constitution of the Indian nation, made evident in the medical notions of the body put forth in the guides. Terms like *Suyogy Santan*, superior offspring, in prescriptive tracts on 'Marital Relations', that is sex, became the normative means of characterizing biological belonging. Instructions for the preparation of the body before intercourse, the appropriate times during the menstrual cycle to have intercourse and, most importantly, the appropriate coupling of individuals were central to these discussions. One vaid wrote in the popular publication *Chand* that

> If [a man] lacks the control to be overpowered by sexual desire like an animal, and thinks he shows power by increasing the number of children every year, he is putting his responsibility on the head of the society, and is a criminal in the eyes of society. Besides those people, these also are not doing their duty: those who are themselves sick, without limbs, and with low semen, and though being like that get married, and give birth to weak children, therein making society crippled and increasing the burden of the world. Because of this rule, our ancestors could give birth to offspring who were world famous, and brought fame and pride to their race (kul)...those who don't keep these rules will always be ill.[31]

Other authors were more explicit in their categorization of social burden: the poor, the illiterate, the uneducated were identified as groups that need not be regenerated. However, whereas discussions of Islamic bodies focused on its separateness, these discussions offered up instead a prescriptive approach to solving the problem of class- or caste-based difference. In the same way that physically ill Hindu bodies could be cured by Ayurvedic remedies, those who were socially ill could remedy the situation by placing national concerns before their own desire.

## Gender and indigeneity: Women and indigenous medical discourse

The early twentieth century saw the emergence of a series of domestic guides called *grhlakshmis*. The titles mostly made reference to the home; for example, the journal *Grhini* was not the only publication to play upon the term *grha*, or homestead, linguistically incorporating the woman into the functioning of the home. Indeed from the later nineteenth century a wider genre of domestic guides in North India was collectively known as *grhlakshmi*. The content of women's journals reflected their titles – every possible sector of the domestic sphere, ranging from the daily living routine, to the length of time needed to cook *dal*, to the most fertile times of a woman's reproductive cycle. Mary

Hancock has argued that *grhlakshmi* journals held more closely than did the school textbooks to the goals of the state-sanctioned 'home science' being formally taught to girls. Textbooks were often vague, confusing or irrelevant, presumably because the authors of the guides had little knowledge of the domestic setting they attempted to address in their work. For Hancock, they were intended to impose a British domesticity upon an Indian setting, and the obscurity of their instructions can be understood as an intentional failing. Conversely, the *grhlakshmi* guides and pamphlets were precise and were informed *solely* by the indigenous domestic environment.[32]

This points to the social control that women both exerted and laboured under as writers, readers and consumers of popular culture. Distanced from the waves of both scandal and success associated with general Hindi-language newspapers, the women's publications relied on the growing consumer base of middle-class, literate women who subscribed to the ideas about family life, morality and social participation contained within the publications. Furthermore, the prominence, popularity and content of these publications demonstrate that the women who read them were vigilant over the content, sending frequent letters to the editor and contributing articles of their own. The emergence of women as purveyors of print culture within the public sphere is linked with both the rise in women's education and the development of popular nationalism. While the increase in education helped women to accumulate the tools needed to participate in the public sphere, the nationalist politic helped them shape the parameters of their agenda.

It is unsurprising that the radical movement of women from the 'home' into the 'world' did not lead to a radical reinvention of the role of women in popular culture or in the national politic: the image created was of the traditional and dutiful wife and mother who put home and hearth first, and whose duty to the nation was to maintain an upstanding household. From the 1870s on, the idea of the *Bharat Mata*, the Indian mother goddess, had been an integral part of the cultural resistance of colonial rule, and it is this image that was upheld in the popular press.[33] However, the anxiety surrounding the maintenance and protection of the iconic Mother India suggests that this symbol was not as straightforward as it seemed, and itself represented different understandings of community and nationhood. Similarly, despite the limited potential encompassed by the idea of the Mother goddess, women writers found ways to exploit the social currency of this fixed image. Assuming that it was in the home that the future leaders of the modern Indian nation would develop, the authors of these publications took it upon themselves to challenge the social, cultural and, at times, political norms in North India by using the discussion of family life to put forth a woman-centred ideal of what the Indian nation could be. Although they worked within the arguably limited framework of a controlled, submissive femininity, the ideas they put forth attempted a reconstruction of what the Indian wife and mother could do – and could become.

Within this paradigm, men had the space to become heroes as protectors of the feminine motherland; as mobile members of a global system, men could venture out into the 'world' as both representatives and protectors of the feminine 'home'. Indeed, men played a role in the publication of women's magazines; often the publishing team consisted of husband and wife, while some of the journals were funded by male-run institutions, like the Arya Samaj grants to journals like *Panchal Pandita*.[34] Men of the household might also hold sway over what the women of the household chose to read. Historians have generally used male-authored examples of literature discussing the role of the male in the domestic sphere to shed light on the oppression of Indian women. However, it is in the idea of protecting the home that flaws emerge in the logic of this argument; while the idea of the male protection of the domestic sphere might have had some social resonance, the constitution of the family life under protection was supervised by their womenfolk. Ultimately, this bestowed upon the figure of the housewife a certain significance that remains unrecognized in a paradigm that explores the structure of the home without considering its construction. While men protected the idea of the 'home', the women who ran the household were ultimately responsible for determining what happened within it.

Furthermore, the centrality of domestic life to larger debates about nationalism bestowed upon the housewife a degree of power that cannot be discounted. As Partha Chatterjee has argued, 'The world was where the European power had challenged the non-European peoples, and, by virtue of its superior material culture, had subjugated them. But it had failed to colonise the inner, essential, identity of the East which lay in its distinctive, and superior, spiritual culture.'[35] For Chatterjee, Indian tradition was thought to have been preserved in the home. However, the significance of these publications reached beyond specifically gendered readings of authorship and audience, giving them value simply as Hindi-language material circulating in the wider public sphere. If anything, the writing on domestic life provided women with an entrance into the public sphere, which subsequently changed the way in which that sphere was constructed.

While couched in the language of morality and duty, the focus on the home reflected the 'traditional' medical community's attempt to incorporate the innovation of modern medical sciences into a revamped 'indigenous' tradition. While the authors made no claims (except in very few instances) to having any sort of Ayurvedic authority, they would often deploy the term 'Ayurveda' to describe their pronouncements. Ultimately, their claim to having an intrinsic sense of how the Indian body worked picked up on the same themes circulating in the revamped Ayurvedic publishing world. At the same time Ayurvedic medicine and knowledge was already a part of household life. The popular discussion of Ayurveda in the public sphere worked to underline its importance and to bring it to the forefront of the domestic experience. Ayurveda was about living in the body and celebrating the balanced functioning of its many parts so that embodied experiences could reach their full potential. Fundamentally, the claim to an understanding of what made the *Indian* – read in this instance as Hindu, twice-born and of an appropriate class background – body work is what tied these discussions to the more formal genre of writing about Ayurveda.

### The emergence of the reproductive body as genre

At the same time, the emergence of discussions about medicine that were accessible to lay people facilitated the opening up of some of the more taboo topics that concerned the nation. The most innovative theme of the *grhlakshmi*-style publications was the focus on sexuality and reproduction. While discussions of cooking, cleaning, child-minding and husband-pleasing came as no surprise, the focus on reproduction, from conception to post-natal care, took a huge leap away from earlier nineteenth century attempts to educate women. Instead, these discussions addressed the cultural politics of Indian motherhood. Concern about women's health had been part of imperial state planning of health care and sanitation, and reproductive issues in particular surfaced in debates on the Hindu Widow Remarriage Act of 1856, the Female Infanticide Act of 1870 and the long-running issue of the age of consent and child marriage.[36] From the Indian perspective, however, specific knowledge of the reproductive body remained generally informal. However, from the 1910s on, overt and more subtle discussions of sex, sexuality and reproduction emerged in various spheres of Hindi writing. The main feature amongst a diversity of opinion was the language of science that proliferated and the perceived service done to the nation through these discussions. As long as the discussions were couched in these terms they could go relatively uncontested in the public sphere.[37]

A good example of this is found in the language employed to educate modern Hindus in the art of appropriate desire. As Sarah Hodges has made evident, one of the key facets of becoming modern in the context of family life was the privileging of the nuclear family, in which the husband and wife make decisions about the household, over the traditional joint family.[38] This line of thought made implicit comments on the backwardness of the joint family structure, in which authority rested on age and status, and not education or civic standing. The stated goal of this 'decision-making' was the rearing of a family of *suyogy santan* (superior offspring) who would serve the nation, and discussions of sexuality always conformed to this end-goal.[39] However, the processes by which this goal was attained were also given space in the pages of this genre of writing. The prescriptive language of science facilitated a foray into the world of sexuality never before allowed to circulate in polite society. For instance, one author had this advice to offer her audience:

> After bathing with sandalwood, applying hari to the body, putting on perfume and having eaten strengthening substance, and wearing clothes of saffron and perfume of musk deer and camphor, or what's available at the time, and having eaten paan, a healthy man who has extreme desire for producing a son and is happy in the love of his dear wife, when he has a strong desire, he should go to his wife. He should adorn his wife with jewels; there should be the desire in both to have intercourse to produce a son.[40]

While the actual curative value of these substances for this purpose is specious at best, the prescriptive language, written in a guide to health, and coupled with the distinctly indigenous nature of the ingredients, leant these instructions scientific authority. The mention of the desire to produce a son justified a foray into the illicit acts implied in the encounter. Finally, the mention of a saffron shirt, widely recognized within the Hindu tradition as an auspicious colour, situated desire within a distinctly communal context. Science, in this instance, responded to the social demands made of it.

Another important theme in the conversations was the rearing of healthy children. One of the most interesting things to note about the new movement in publishing about pregnancy and health care was the conception of audience held by those who wrote about it. Whereas nineteenth century guides to childbirth, though few and far between, were written for midwives and for lady doctors, the new writing on pregnancy and childbirth was directed at the same women who consumed the printed knowledge about birth control and childrearing, namely the middle-class mother. Traditionally, the midwife would have been the authority on popular understandings of the reproductive systems. However, this relationship had been undermined by the involvement of the colonial state in the regulation of reproduction, as we saw earlier.[41] This led to training, licensing and professionalization of *dais* from the 1880s, provided they trained at one of the courses set up to educate them in the ways of modern medical science.[42] This paralleled the regulation of other groups who held knowledge useful to the imperial state, as hakims, vaids and pandits came to be acknowledged as professionals once they became associated with emerging institutional bodies.

Therefore, the emergence of a public discussion of reproductive health in the printed form, written in vernacular languages and intended for popular consumption, was innovative. Older guides for midwives had been laid out in very specific ways, detailing the step-by-step process of birthing and focusing on what a birthing attendant could do to help a labouring mother.[43] While there was some concern about the overall gestation period, the majority of the information contained within the nineteenth-century midwifery guides focused on the birth and the potential problems that might have occurred at the time of delivery. For instance, Yashoda Devi, in her *Dampati Arogyata Jeevanshastra (Guide for a Health Married Life)*, detailed the development of the foetus in each month in a section called the 'Cures for Illness in the Womb'.[44] In the section she described the various herbs and plants that should be imbibed in each month to ensure the healthy development of the child. The birthing process is not ignored but is discussed in medical terms, explaining the way in which the baby turns in the womb and descends through the vagina. She also provided 20 pages worth of visual diagram depicting the different stages of a birth, beginning with the turning of the child and ending with its final descent.

Other guides instructed mothers on the uses of their bodies for childrearing after birth and how to make judgements about the health of their children. Guides entitled *Shishu Palan* (Illnesses in childhood) and *Hindu Mata* (Hindu Mother) were amongst the most popular publications of the day, sold cheaply but published in huge print runs.[45] A very popular 1918 guide called *Dudh-Chikitsa (The Curative Aspects of Milk)*, written by Ramnarayan Sharma, explained the benefits of milk to ensure a child's healthy development – but also to ensure the mother's recovery.

The book begins with the author's reference to the work of a *videshi mahila*, a foreign woman called Edith Wheeler Wilcox, who claimed that a lack of milk in children led to *hrudayse sambandh raknevale rogonko*, the translation of which was provided in parenthesis as 'organic heart trouble', and milk could also rid a body of cancer.[46] However, in the next line, the author tempers this foreign knowledge of the body with the reiteration of the appropriateness of native medical regimes for native bodies, claiming that 'a body will also be free of illness if a plain diet is followed, according to the *panchjanevale*, the Ayurvedic fiveway path, and if spicy foods like *mirch masala* that have a sharp edge are avoided'.[47] In this text, we see the coming together of a variety of themes that framed the discussion of female-centred medical discussions within the larger context of Ayurvedic and nationalist discourses. First and foremost, the author tempers his discussion of female reproduction within the context of the nation, using the health of the national child to speak about the potential ill health of the mother. Secondly, health edicts that were foreign in nature were made appropriate through their incorporation into a wider spectre of health care based around a properly indigenous system.

There were sceptics amongst these female authors and those who believed firmly in the advances of biomedicine and wished to leave the ancient cures behind. One very popular author called Sushila Devi grew exasperated with the focus on the natural, indigenous cures that were consistently pushed in these guides. A loyal follower of Marie Stopes, Sushila Devi wrote a tract entitled *Dampati Jivan*, or Married Life (clearly borrowing from Stopes' masterpiece *Married Love*), in which she introduced readers to foreign, innovative technologies like condoms and pessaries, advocating their use in the fight to build a strong nation. Writing in 1931, she challenged her readers on their veneration of the 'natural': 'If you are only in favour of natural things, why don't you go on foot from Bombay to Calcutta? During the monsoon, why do you go out with an umbrella? Why do you go to the doctor when you have a fever and why does it stop when you take quinine?'[48] Despite the deviation in form, the themes that Sushila deployed, coupled with her commitment to nation-building through appropriate and responsible reproductive techniques, situated her firmly within the norms of discourse that pervaded medical writing in this period. Ultimately, it was the focus on nation-building through appropriately 'indigenous' means, even from the perspective of its critics, that allowed for profound discussions about gender to be staged.

## Indigeneity beyond Ayurveda

Several authors who wrote about medicine outside of the context of the sphere of Ayurvedic writing and practice drew upon this discourse to provide a context for their own writing. This is particularly true of discussions initiated by female authors. In particular, cooking and female health issues drew extensively from the premises as well as the cultural authority given to Ayurveda as an indigenous system. The idea of indigeneity, espoused by vaids in their rooting of modern practice within an ancient tradition, provided a framework for women to expand their discussions of the 'timeless' practices of the domestic sphere. The proliferation of the idea of indigeneity rooted in scientific evidence created a space for their discussions to permeate within the Hindi public sphere as serious subject matter. In fact, it was the already established discussions that created a space for the intimate details of married love – and the darker side of human sexuality – to be publicly discussed. The self-declared Vaid Yashoda Devi, the most prominent and popular female writer of this period, used Ayurvedic authority to deploy new information about the female body into the public sphere without fear of major reproach.[49]

Yashoda Devi claimed to be a traditionally trained Ayurvedic doctor and writer, trained by her father and then her husband, with whom she ran several private clinics. She had a very long and active publishing career spanning the first four decades of the twentieth century.[50] It is in her writings that we see the coming together of a discussion of health and health care articulated in language and a writing style that appealed to Hindu women of the emerging middle class; in essence, she made Ayurveda resonate in new ways for women. She was unusually well able to capitalize on both the complicated politics of Hindi-language scientific writing and the boom in publishing for women, and turned her writings into a successful venture that fitted both agendas. On the one hand, her insistence on *shuddh* (pure) Hindi and its scientific focus helped her to lay claim to medical writing in the Sanskritic 'tradition'. On the other, the centrality to her arguments of the domestic sphere and the 'appropriate' role of women made her extremely accessible to middle-class female consumers of popular printed materials. She intended her guides to be a reflection of what her society could be, but her vision of the future was steeped in, and guided by, memorializations and resurrections of the past. Her attempts to establish continuity between past, present and future, and so to stabilize an 'essential' Hindu identity, were directly in line with the arguments of other vaids.

However, some of her interpretations of the appropriate role of women, both in the family and in building a new India up out of the next generation of families, were distinctively her own.

Women had traditionally disseminated knowledge about women's bodies to other women, and it was upon this cultural norm that Yashoda Devi could capitalize.[51] While treating men was not something a female practitioner of 'traditional' medical knowledge could yet do, addressing women's health was a project that would find both cultural and political currency in the context of the new concerns about the health of the 'modern' Indian nation.[52] However, it was her claim to a more neutral, degendered identity as a vaid that allowed for her writing to be consulted outside of the scope of the sphere of women's writing. She included her own glossing of Sanskrit edicts, continually claiming that her ideas emerged from the 'shastras'. Using the structure of the Ayurvedic textbook, Yashoda Devi expounded upon medical topics from a woman's perspective, rooting Ayurvedic edicts within the context of the domestic sphere and charging Hindu women with responsibility for the health of their families. By couching her claims about difference and belonging in medicalized language, Yashoda Devi attempted to invoke the reworked authority of Ayurveda in this period.

Like other vaids, and indeed other female authors, she invoked discussions about the constitution of the nation, thus placing her firmly within the most important debates of this period. She too concluded that Unani was harmful to Hindus and more acceptable for Muslim bodies.[53] She also attempted to pathologize the practice of Purdah, claiming that it benefited the health of Muslim women, but also implying that it protected Muslim women from their overly sexualized menfolk. Yashoda Devi's major concern, however, was with the improvement of the extant nation, and she conformed, as did the majority of Ayurvedic writers, to a eugenic ideology that was fundamentally inward-looking and disinterested in the disciplining of other groups.[54] Yashoda Devi's central concern was to produce strong, worthy citizens, who would make up a *suyogy* (competent) nation. This fuelled her interest in health, and it was the ferocity of this belief that caused her to become so active a writer and publisher. Her emphasis on the glories of the past bore the most fruit: in her writing, problems like overpopulation were cast as thoroughly modern issues, in contrast with the ancient past, when sages were wise and people were pious. Educating Indians in traditional practices, therefore, could solve the population problem. In particular, followers of the Arya Samaj would have been familiar with this argument of 'improvement'. As with her domestic guides, the middle classes needed to be educated in appropriate reproduction, lending credence to Sanjay Joshi's suggestion that the North Indian middle class was more a project than an established entity.[55] Moreover, in her discussion both on the preparation of food and on the delivery of children, working-class labourers emerged as the 'other' in relation to which Yashoda Devi and others like her constructed their position in the economic hierarchy of North Indian society; to consider the significance of food itself instead of its preparation, or to consider the particulars of conception and gestation above and beyond the process of delivery, was to demonstrate prosperity sufficient to allow families to concentrate on social significance rather than survival.

Yashoda Devi's articulation of class distinctiveness was not explicitly socially divisive, as she stressed the improvement of the individual over the control of the population. She argued that ignorant parents produce ignorant offspring, and that 'women who have too many babies bring ignorance upon themselves'. In essence, Yashoda Devi's work symbolizes an endpoint to this approach to the history of Ayurveda. By the 1930s, the idea of the Ayurvedic text was not only sufficiently diffused but also sufficiently established that it could give way to a larger discussion of indigeneity and belonging rooted in the language of science and medicine.

## Consuming indigeneity: Medical advertising

Advertisements for medicines and other products related to the body provide perhaps the most interesting ground for investigating the more subtle meaning of 'health' and the body in society. Advertising was a part of every genre, and played a similar economic role in the history of publishing in North India, and was policed more by the tastes and desires of the readership than by the agenda of the magazine in question. For instance, as Charu Gupta has shown, respectable newspapers like *Vartman* and *Chand* might closely regulate the content of the articles on its pages for inappropriate sexual behaviour, to the point of promoting discussions of sexuality as appropriate only within the context of reproduction, yet routinely sell advertising space to those offering aphrodisiacs or cures for venereal diseases.[56] As such, they provided a reflection of the other extreme: Sexual desire outside of the confines of marriage was something against which the middle class railed on moral grounds, while the products sold to aid desire (or remedy its ill effects) were consumed in great quantities. Advertisements then, be they for patent medicines, Ayurvedic remedies or for improved sexual activity, represent, in some respects, a counter-sphere to the discussions of medicine and the body more formally debated in the form of articles, chapters, pamphlets and books.

Medical advertisements also reflect the popularity of certain concepts at certain times and to certain audiences. Though the advertisements deployed in the pages of magazines, on the backs of books and in the vernacular newspapers were certainly complex, they also bore a simplicity necessarily lacking in literary debates about medicine and society. The purpose of an advertisement was, fundamentally, to sell a product. Advertisements were also key to the education of desire, to the deployment of nationalism, to the cultivation of consumer taste and to the cementation of cultural values within the middle class. However, all of these were secondary outcomes and were fundamentally subject to the vending of goods to medical consumers.

Advertisements, by the first decades of the twentieth century, could be found in all printed materials. The most authoritative guides to Ayurveda, authored by pandits still working within that tradition, bore adverts for other books in the series. Ayurvedic practitioners sold both goods and printed materials in their pamphlets and books. Women's journals were full of ads for 'health products', ranging from products to beautify the reader to consumer goods they might want to purchase. However, two major product groups, with two very different modes of articulating value and selling the product can be detected in this genre. The first is the product targeting the power (or lack thereof) of Indian men, often having to do with the sapping of his virility through disease. The second are the ads aimed for information about babies and the family, usually through books or through enriched medical products (or food substances).

### The diseases of men

The first category of advertising has been of interest to historians and is the most innovative and striking endeavour of the period. From the 1910s on, advertisements for diseases specifically targeting the reproductive systems of Indian men started to appear. Advertisements, however, took a different approach, reinserting the idea of the patient into the equation by transforming him into the consumer. The big gender shift in medical discourse had been the inclusion of the female body, designated through her reproductive tract, into discussions of medicine in the twentieth century. However, the dominant 'neutral' figure deployed in medical theory was still male, and Ayurvedic textbooks routinely deployed male skeletons, corpses and forms in their pages when describing the nature of disease and its cure. However, the demands of a consumer-driven market for health products required a clear delineation of the target consumer, and it is in this instance that we see the specification of the male body, mostly through the moral category of 'strength' and through the physical category of reproduction, emerge in medical discourse.

The most common cure offered was for gonorrhoea, often translated as *suzak*, but almost always transliterated in Hindi. Advertisements offered remedies for the 'killer' disease which threatened, as one advertisement put it, the 'health of the nation'. Gonorrhoea had been a concern for imperial administrators from the 1850s (if not before – it was only in the 1850s that legislations around it emerged) and also appeared in sections of books that dealt with the reproductive health of the couple. However, in both approaches, the male figure was never explicitly described: colonial and later public health policy had worked to located the cause of disease in the bodies of the prostitutes from whom soldiers allegedly caught it; in health guides, especially those aimed at women, gonorrhoea remained a purely medical condition, and readers were given instructions for various concoctions that might eradicate it.[57] Gone was any moral judgement of the possibly transgressive act through which one had contracted the disease, and in its stead was a champion of his sexuality and virility.

Take, for example, an advertisement that appeared in the popular women's journal *Madhuri* in 1940.[58] This journal was targeted at women, though men did write columns, and presumably read them as well. In the advertisement, a large-muscled man points one arm forward and grasps a cipher in the other, which bears the word Gono-Killer, in English, in front of his crotch. At his feet writhes a snake with the word 'GONORRHOEA' written across its flank. An arrow links a box of text at the right of the page to the advertising logo, explaining that the product for sale was called Gono-Killer and that it targeted the disease. The image is very much centred at the man's crotch, where the advertising logo lies, and where the arrow points, not so subtly reiterating that the subject of concern here is the penis. Interesting, also is the figure of the snake. It is not clear if the snake is meant to embody gonorrhoea itself, or if it's meant to represent an infected penis. Regardless, the image of the snake reinforces the reader's focus on the reproductive organs of the male.

At the same time, the extent to which the figure is Indianized allows for the ad to offer more profound commentary that transcends the realm of sexuality. The figure is mostly naked, wearing only a loin-cloth and a sash across his chest. While the sash might be taken as being potentially analogous to the sacred thread worn by Brahmins, and the loincloth is similar to that worn by a Sadhu, the visual cues that might symbolize the sacred or divine embodiment of spirit in a holy man are notably absent. The figure instead looks like a wrestler, an icon which was being reworked in this period as a figure of 'traditional' representation of the virility and strength of Hindu men.[59] The figure, then, can be understood as an embodiment of national strength, gained through indigenous practices relating to physical fitness. The strength of the body was therefore not only a timeless one, located in ancient cultural practices, but also the basis upon which the nation would become independent. At the same time, while the deployment of an English name for the disease and its cure, but matched with a Sanskritization of the term, created an ambivalence around the nature of the disease, was it foreign evil or was it local? Regardless, the cure, though labelled in English, was certainly local, and the body that was to be saved was undoubtedly a national one.

### Healthy babies, appropriate families

The second most popular form of advertising related to the health of families, represented by the health of the children. The nuclear family, consisting of mother, father and child, at once served as both microcosm for the nation and also as its ideal, pushing an image of family that was far from the context in which most people lived. Advertisements provided the perfect forum for this sort of imagining of the family: they were reliant on image, and they didn't have to engage with anything short of the reality they attempted to create. The affective states invoked in literary discussions about the family did not translate into the short, perfunctory and strategic statements used in advertising. Ambivalence, instead, was encoded in different ways, namely through the deployment of certain sorts of images and the claims made about nationalism, race and identity. However, the overall consensus about the importance of the child and, by extension, the family to the nationalist project allowed for the introduction of new themes and new products through the lens of advertising.

One example of this genre is an advertisement for a product to aid in the conception of children. The placement of the text was evocative: the first large print line read 'medicine with which to conceive offspring'; the second line had a box around the word 'baby'; the third line employed the emphatic Hindi tense to assure readers 'you'll be able to have one'.[60] The smaller print around the boxed term 'baby' empha-sizes that this medicine is to help women conceive (*har stree ko baccha ho sakta hai*), and the description of the product in the accompanying para-graph explains that the product cures infertility in women, and can be obtained through 'Lady Doctor Zenana Dawakhana' in Delhi, all inscrib-ing the advertisement as one fundamentally targeting the special needs of women with regard to reproduction. At the same time, the image of the happy family, with crying babes in arms, reinforced the idea that the woman is not doing her duty to her family – and thus to the nation – if she lets her reproductive problems continue unchecked.

Interesting also is the ambivalent racialization of the family under concern. In this picture, it is not clear if the family is Indian or not. The individuals in it differ from other representations of adult Indians, in which women are adorned with *bindis* and wear saris, and in which men might sport aspects of traditional dress even if their style of adorn-ment is westernized. However, the means by which the medicine can be obtained – through the Lady Doctor Zenana Pharmacy of Delhi – invoke a long history of encounter between allopathic medical infrastructures and the Indian woman. Though this medicine might allow the Indian woman to 'modernize' herself and her family through her consumption of medicine, the modernity into which she enters.

Another advertisement that deployed similarly ambivalent notions of race, encounter and the universality of the family used the image of a white child who was racialized as Indian in the text of the advertise-ment. A smiling blonde baby boy sitting on a scale was featured atop this advertisement for vita-milk, an enriched milk substance imported to the subcontinent.[61] Interestingly, the para begins with a historicization of child health-care norms: 'In the ancient times and throughout the land of India, parents used to have to consult a vaid if their children were not growing.' In modern India, they just had to feed the child vita-milk. But what is to be made of the disjuncture between image and text regarding the issue of racialization? Again we see the deployment of an accultur-ated body clearly meant to represent a cultural ideal. Though the text renders him Indian, his image in no way indicates his nationality, as other images of children clearly did.

In the first advertisement, the parents holding the child were clothed in Western dress and were absent of any cultural markers that might make them Indian. Instead, they referenced some default identity, which historians of colonial-era advertising have designated as the white body. I am hesitant to use this model in this context, as the bodies in question are not clearly white either, but they do invoke a form of the family that is clearly not self-consciously Hindu in its representation. Instead, I'd like to suggest that the message of the advertisement is the invocation of an affective state of joy at parenting, and the employment of a 'universal' image of a family, devoid of cultural markers. The emphasis on the 'universal' family, as made evident in the culturally neutral – and most probably Caucasian – images of babies and parents, works to promote the imagined nuclear, modern family as an ideal that transcends the specificities of race and location, a joy for anyone participant regardless of race or ethnicity. However, it was the iteration of ideals of the nation deployed in other contexts that made baby-making and child-rearing such an important ideal in the first place – important enough, in fact, that it could transcend the image of the nation even as it worked to promote it.

The deployment of notions of Ayurveda and of biomedicine was common in these advertisements, ostensibly both covering bases of authoritative control and making evident the tension between the two traditions that existed. A second advert placed by Lady Doctor Zenana Dawakhana advertised a white water for women (*stree ko shvet pani*), promising to cure their colds in three days, and insisting both upon the pandit-approved nature of the medicine and of its ability to stand up to the rigid testing standards of the *naye yug*, the new age.

\*   \*   \*   \*

Ayurvedic authors claimed a nascent authority over the realm of medical knowledge, leveraging for authority in new linguistic and textual forms, but never ceding their authority as traditional experts; individual female writers both railed against and happily appropriated the tradition doled out, but always cast themselves as masters of the domestic domain, as Yashoda Devi made evident. Finally, medical companies, like the Lady Doctor Zenana Dawakhana, offered technologies and techniques devised by their own practitioners, claiming the role as alternative experts in a world of consumer choice.

At the turn of the century, pandits and biomedical doctors were still the arbiters of medical knowledge; similarly, the negotiation of domestic practices was disconnected from public life and civic identity. The emergence of a discourse of indigeneity that recognized highly sophisticated and highly informal knowledge about the body as the two equally important poles of an authentic national identity helped to erase these rigid boundaries around knowledge production. Key to the project was the articulation of these ideas in a new space and in new ways, which was found in the Hindi public sphere in all of its many manifestations.

However, the eradication of the author from the discourse by the 1940s makes evident the extent to which he or she had occupied a huge position of power in earlier writings. By articulating these ideas in new, modern and nationalist idioms, authors could reconfigure their authority by reinventing the traditional roles they occupied. Pandits could move beyond the realm of the sacred and communicate resonant ideas to a new public; housewives could claim new authority outside of the home by structuring their knowledge as vital to the rearing of the family. In either case, the claim to be the disseminator of information crucial to the constitution of the nation ultimately paved over any divergences in theme or practice and provided a gravitas that made even the most arcane or bizarre practice or ritual sound at least somewhat profound.

Also linking these two divergent processes was the subjective experience of author as nationalist subject. Authors were Indians first, in body and in politic, and vaids or authors or entrepreneurs second. Their subjective experience of political, historical, social and cultural events clearly differentiated them from those who had determined medical principles in the past. These new authors were very much in the world, in both language and sentiment – not in the ashram or the temple, nor working in a language at a remove from the population. The knowledge deployed was common, resonant and easily accessible. Ultimately, in writing these guides, medical authors came to take on the same characteristics. The self-conscious representation of themselves as part of a collective attempt to return to an 'authentically' Indian way of living, which had been corrupted by various processes, inserted them into a new sort of hierarchy. Once it was resolved that social practice had to change, they had the authority, constructed through the authority of tradition combined with a fluency in contemporary social values, to provide the information that would make this change a reality.

# 4
# Ayurveda's Dyarchic Moment, 1920–1935

As we saw in Chapters 2 and 3, the consolidation of Ayurveda as a singular, unified system of medicine relevant to the lives of 'modern' Indians happened, in part, through the identification of the practitioner as the arbiter of the tradition. The GOI's Medical Department had turned to the experiences of the practitioner when determining policy matters in which the Indigenous Systems of Medicine were implicit. Similarly, the practitioner had emerged as the new Ayurvedic 'expert' in the Hindi public sphere, replacing the Brahman Vaid. This chapter explores the ways in which the new authority granted to practitioners was further entrenched through its incorporation into the medical infrastructure of the United Provinces in the 1920s and 1930s, an act which fundamentally transformed the Indigenous Systems of Medicine by recasting them as modern, state-sanctioned structures of bureaucracy and service provision. Vaids and hakims were important subjects of provincial medical reform and became the vehicles through which the Indigenous Systems of Medicine were modernized.

Though the episodes discussed in Chapter 2 reveal the extent to which the Indigenous Systems of Medicine had gained some recognition within the framework of the imperial state (mostly through the actions of individual practitioners), their adoption as an important part of the provincial medical infrastructure marked a major shift in colonial medical logic. Throughout the nineteenth and early decades of the twentieth centuries, the imperial government had continued to think of 'Indigenous Medicine' as a unified category, framing the systems as those that entailed medical practices that were not 'scientific', and criticized the lack of standardization amongst indigenous medical practitioners.[1] However, the expansion of the bureaucracy following the Montagu–Chelmsford reforms had resulted in the incorporation of a new class

of 'native' servants into the provincial administration, who were subject to the same rules that had governed their European counterparts.
The Province, newly responsible for its medical responsibilities to both
administrators and subjects, could no longer afford the expensive
biomedical bureaucratic measures it had used in the past to certify the
(ill) health of its bureaucrats. The provincial government consequently
decided in 1923 that lower-class officials could see the practitioner they
usually consulted in order to get a certificate of ill health.

However, this was not the simple shift that the government seemed to
have envisioned. A variety of questions arose regarding its implementation: Who were these practitioners? What made them qualified? Were
they quacks – what determined if they were or were not? So began
the complicated process of trying to gauge what exactly indigenous
medicine was, and soon after followed the regulation, legislation and
professionalization of the indigenous medical systems in the UP for the
purposes of government. The government established a Board of Indian
Medicine that would advise on matters concerning Ayurvedic and Unani
medicine and that would act as a liaison with practitioners. The regulation of indigenous medical practice was necessary if government was
to maintain control over the Indigenous Medical Systems, which had
become crucial to the bureaucratic processes of the reformed provincial
government.

The decisions made by newly elected Indian officials in the earlier
part of the period were limited to the reform of aspects of the tradition
that best suited governmental needs, focused on transforming the traditional practitioner into a modern political subject, and did not attempt
an overhaul of the systems themselves. In contrast, the decisions made
in the late 1930s took advantage of the new legislative power of governmental institutions to expand the scope of educational institutions
and medical care facilities, while pushing for further reform of the practitioner's position in society and government. At the same time, where
politicians in the 1920s and 1930s envisioned the reform of the indigenous medical systems as crucial to the maintenance of the colonial
state, the Congress-led government of the post-1937 period saw the radical development of the indigenous medical systems as the key towards
creating health policy for an independent state.

This chapter focuses on the articulation of communal politics and
affiliation in the context of the development of the indigenous medical systems. The expansion of the Indian participation in governance,
which had increased dramatically after the 1935 India Act, had been
crucial in developing the political articulation of communal identity, as

it entrenched and expanded separate electorates for Muslims, landlords and scheduled caste candidates, thus recasting the political process as one that would be representative of the community and not the nation.[2] This chapter explores the ramifications of the politicization of communal consciousness in organizing indigenous medicine in this period. The professionalization agenda of the 1920s and 1930s begins to draw sharper lines between vaids and hakims, while giving them equal status: the Board of Indian medicine was constituted of an equal number of vaids and hakims; however, the insistence on balance reflects the implicit belief, circulating in the public sphere, that Ayurveda was for Hindus and Unani was for Muslims. Ultimately, the addition of a communal element to the logic of regulation, certification and legislation, as we shall see in Chapter 5, was something that vaids could exploit to protect their newfound power. Later in the period, the government began to distinguish between Ayurveda and Unani, relying on communal affiliations to drive new medical policy. Congress politicians began to look at Indian sciences on their own terms and as a reflection of Indian culture, and so adopted a distinction between Ayurveda and Unani, despite their continued rhetoric of unity. In this context, Ayurveda was the dominant traditional system and vaids were the appropriate indigenous practitioners.

This chapter also offers insight into new trends in planning that arose in the interwar period sandwiched between the Montagu–Chelmsford reforms and the India Acts of 1919 and 1935, both of which worked to transform the working of politics on the ground. The 1919 reforms saw the advent of dyarchic rule in the subcontinent, in which the rights and responsibilities related to the governance of non-essential matters of the centralized state in Delhi were shifted to newly empowered provincial governments. This system, known as dyarchy, lasted until the transformative India Act of 1935 radically transformed the spectrum of governance, increasing the suffrage to 35 million, and ushering in a proper party system based on fully engaged electoral practices. This period, written off by historians as being one of grand-scale political failure due to the inadequate logics and of governance that dyarchy represented (as well as the paltry reach of indigenous politicians at the centre), is, however, also one of great possibility. A view from the ground – in this case, 1920s and 1930s' United Provinces – recasts this take on the evolution of politics by highlighting the immense creativity and possibility with which local politicians were able to govern. Left with huge tasks, little support and a lot of autonomy over formally mundane (but truly crucial) portfolios, the decisions made and policy created during this period of late colonial rule offer a tangible break from the colonial order of things.

## Power and governance in the interwar period

The advent of a dyarchic structure of governance following the implementation of the Montagu–Chelmsford reforms of 1919 was intended to grant a small measure of self-government to Indians. The narrative around dyarchy stresses the sacrifice of Indians and understands the meagre gains made through dyarchy as payment for services rendered during the war. However, Lionel Curtis, the renowned Imperial federalist who worked out the logic of dyarchy in India, envisioned a different trajectory. For Curtis, self-government was a reflection of the way in which Indian politicians were already mobilized into politically sophisticated forms, evidenced by their own advocacy for majority representation within legislative councils, as well as their decision to formally back the war.[3] Self-government for Curtis was conceptualized in steps, introduced gradually, until Dominion status could be conferred. Instead, 'provincial electorates through legislatures and ministers of their own [could] be made clearly responsible for certain functions of government to begin with, leaving all others in the hands of executives responsible at present to the Government of India and the Secretary of State'.[4] While Curtis was arguably more optimistic about the swift trajectory towards full self-government than were his compatriots, his plans for dyarchic governance were implemented from 1919.

The aim of the reforms was to begin to introduce responsible government in India through the introduction of the democratic principle into the executive branch. To begin with, the franchise was extended, though only to 3%–4% of the population.[5] Moreover, Indians elected to the provincial legislature could now be appointed to the role of minister and could oversee the administration of governance within the provinces of British India. Most significantly, the reforms proposed a division of political portfolios into reserved and unreserved topics, the latter of which would be transferred to the provinces for local administration. The reserved topics, unsurprisingly, took up the large-scale political work of Empire, including the rule of law order (through the military, the police and the legal system), the foreign affairs of state and any files related to the economy; unreserved files took up local self-government, education, public works and agriculture.

The electoral reforms merely extended the system already put in place ten years earlier following the Morley–Minto reforms of 1909 and were, ostensibly, the next stage in a series of extension of the franchise. Far more significant was the creation of elected ministerial posts. The Morley–Minto reforms had merely allowed Indians to elect non-official members of provincial boards; after the Montagu–Chelmsford reforms, elections created political figures with the opportunity to implement policy. As such, dyarchy is perhaps better represented as the addition of the onus of responsibility for these services along with others already encompassed within the scope of provincial political and civic responsibility. Ostensibly, it was not so much a transfer as it was a reinvention of a new category: medical policy at the centre of British India had been primarily concerned with large-scale public health epidemics, but had little to say about the every-day lives of Indians. However, the expectation of the provinces was that the medical policy they created would mean something more profound. Though the reforms were represented at the time by Lionel Curtis and others as a devolution of 'power' from the centre to the provinces, the dyarchic structure of governance is perhaps better understood as the creation of new channels of political accountability.

Dyarchy has been mainly considered within the context of the electoral – and hence political – power that it bestowed (or failed to bestow) upon colonized subjects. It has thus been dismissed by historians as a significant factor in the development of political organization amongst the native populations. Scholars have relegated the Montagu–Chelmsford reforms and the India Act of 1919 to the realm of other token provisions for power with underlying agendas that enforced rather than deflated the colonial dominance of the British, arguing that while it extended the reforms made ten years earlier, it also entrenched the already divisive communal politics, especially in zones like the United Provinces where communal tension was prevalent. Historians of the 'Cambridge School' of Indian history have explored the potential for the expansion of civic identities amongst those able to participate in the system directly, recasting the new as significant to the entrenchment of bourgeois society in the major provinces of the subcontinent.[6] However, recent work on the middle class has complicated this argument: Sanjay Joshi, in his work on the United Provinces, has located the class-formation project within more subtle and immediate social and cultural projects.[7] In a study of health and governmentality in Madras Presidency, Sarah Hodges argues that the Justice Party made use of dyarchic ministries to further the cause of non-Brahmanic social and political reform that was not contained within Congress-style nationalism.[8] Despite these contemporary historiographical advances, the dominant approach has been the dismissal of these phenomena as the least important of three political moments, casting dyarchy as the bridge between the Morley–Minto reforms of 1909, which created the suffrage of Indians, and the Government of India Act of 1935, which vastly increased it.

Stephen Legg has productively framed dyarchy through a model of scalar division, in which the seeming divisions of region, centre and ultimately empire are deconstructed not only to emphasize both the dynamic opportunities of movement across and through them but also to refocus these broad categories on the lives of individual actors.[9] As such, scale is employed to breakdown broad categorization and to allow instead for effects of reference rather than structures or frames, made relevant by the working of networks and other modes of relation or assemblage that reorganize the way in which scholars understand the interconnectedness of actors and phenomena.[10] This way of thinking about the pragmatic and transformative effects of knowledge production about population (and its pragmatic ramifications in policy) is a productive tool with which to explore the history of dyarchy and its effects. The impact of dyarchy within the framework of late empire casts it as a political defeat for elite nationalists (and the populations they purported to represent), an insult that seemed to diminish the sacrifices made by the Indian colony during the First World War. A view from the provinces, however, offers a different picture, one that is illuminated by the notion of scale. While the large-scale portfolios of defense, currency, foreign affairs, policing and even the public health remained the purview of the Raj, provincial self-governance was given a lot of autonomy over localized issues, including health, education, lands, road upkeep and agriculture. In essence, a scalar political reading of dyarchy reveals the ways in which the devolution of non-essential portfolios resulted in the ongoings of everyday life coming under the purview of elected officials. While this had been the case for municipalities from the 1870s on, the upgrade of these issues to matters of provincial import created the potential for the creation of schematic planning that could take into account populations and geopolitical locations in tandem.

Historians have stopped short of considering the further bifurcation of power in this period, namely, the division of both power and allegiance to the colonial state and to the ideal of an independent nation. The evaluation of the dyarchic structure from the perspective of political advancement and the civic responsibility of the elite ignores the more subtle changes that these reforms had enacted, some of which had effects that were surprisingly transformative. The transfer of responsibilities for certain essential services resulted in a re-evaluation of the practices of local government, and, in some cases, a revamping of the institutions and systems of governance employed by the colonial centre. Medicine had been a crucial area of intervention for the imperial government from its ascendancy in 1857 through the First World War; interested parties at the provincial and central levels relied upon medical institutions and institutionalization of medical policy to maintain both financial and social control over their territories, and the health of the Indian public also had broader imperial resonances.[11] While matters pertinent to ensuring the public health of the population against epidemics, famine or other crises of potentially catastrophic proportions remained the responsibility of the centrally controlled Public Health department, the every-day workings of the Medical department were transferred. Recast as a localized issue, the agenda of the newly formed provincial Medical Board now had to represent local concerns. The bureaucratic changes of the dyarchic era had redefined the scope of politics by expanding the demography of political involvement. The Medical Board in the provinces had to respond to a new social and political order and ultimately adapt to the revamped power of the province.

The institutionalization of medicine and medical policy was thus bound up with the politics of the Raj in its provincial manifestation and the new political empowerment of Indian actors. This was, of course, not a new phenomenon: historians of medicine and empire have consistently stressed the importance of the governing of the body to securing imperial power around the world. The re-inscription of the new political landscape with older political paradigms is unsurprising, but this particular manifestation of the intersection of medicine and politics was uncharted territory. Earlier models of medical governmentality were characterized by the conscious attempts to establish the dominance of the Raj over Indian political life, in part through the imposition of a medical logic that would, at times very invasively, permeate the intimate spheres of Indian life.

While historians have debated the reach and the limits of the colonial public health and medical systems, as we saw earlier, the logic of the colonial state clearly relied on the disciplining of the colonial body as part of its attempts to forge authority and maintain power.[12] Of course, as Arnold elegantly puts it 'medicine was too powerful, too authoritative, a species of discourse and praxis to leave to the colonizers alone', and the history of medicine and the body is best characterized when understood within the context of the 'onion layers of resistance, accommodation, participation and appropriation' of medical intervention by

Indian actors.[13] Despite the complex interactions between colonizer and colonized, medical policy under both the EIC and the GOI had at its centre the introduction – and in some instances, the imposition – of 'modern' allopathic medicine to the subcontinent. Even after the development of local self-government from the 1880s on, public health agendas were ultimately ordered from the centre, and at times left to provincial magistrates to implement.[14] Following dyarchy, both aspects of this policy changed: the medical systems central to the project were of the indigenous variety and the newly formed assemblies of the provincial government were the sole arbiters of their dissemination and use.

The provincial government, through the institutions it created to deal with the transfer of responsibility for the medical services, laid out as its primary agenda the professionalization of the Indigenous Medical Systems. This shift was hugely significant: the systems themselves had come under fire for over 150 years and had been represented in the British political, cultural and professional imagination as philosophically outdated and theoretically backwards. However, from 1921, the critique of the system was redirected: the practitioners, their education and training, the institutions in which they practiced and the infrastructure (or lack thereof) to keep them standardized were identified as the causes of the state of disrepair into which they had fallen. The dyarchic agenda and the reformation of politics in the United Provinces took a two-pronged approach that actually mirrored the larger extent of the dyarchic intervention. Firstly, the expansion of political responsibility imbued in the newly elected Indian officials was reflected, as we shall see, in the attempts by the Board of Indian Medicine to discipline indigenous medical practitioners into becoming responsible citizens. Secondly, the expansion of responsibility for medical services was partly addressed by the creation of new institutions, or the revamping of existing ones, that could regulate, organize and make uniform the dissemination of health services.

The reform of the Medical Services thus provides a valuable lens through which to evaluate the culture of politics between the wars, and under a new regime of province-centred power. Moreover, it tells the story of the ways in which Muslims and Hindus were polarized from each other through basic social institutions established on ostensibly 'neutral' grounds. Ultimately, the reform of Ayurveda and Unani as Hindu and Muslim traditions, and the creation of institutions and modes of practice that entrenched the difference between them, represented the move towards new political pre-occupations and new modes of political practice.

## Making medical 'professionals'

The Indigenous Medical Systems became a target of provincial governmental reform in the post-dyarchic period in response to the changing medical needs of government personnel. The attempts to create policy were tentative and limited in scope as Indian politicians negotiated their new roles as political officials with some power but limited political capacity. The newly formed Board of Indian Medicine, a branch of the United Provinces Department of Medicine, embodied this cautious political manoeuvring in its attempts to expand the medical field to accommodate the increased colonial bureaucracy. The Board's primary focus was on the professionalization of the Indigenous Medical Systems, enacted through the reform of the practitioner's status and position in the legal framework of the medical bureaucracy. This approach brought about two sorts of change: firstly, it enabled an expansion of government in a way that was financially feasible by offering the bureaucratic medical services necessary for government protocol at a fraction of the price; secondly, and more importantly, they symbolized the propensity for the modernization of tradition in a way that could serve the interest of the state.

The Board of Indian Medicine was formed in 1921 following the transfer of certain responsibilities, including those of the medical and public health departments, to the provinces after the Montagu–Chelmsford reforms. The Board was established to advise the United Provinces' Medical department on issues pertaining to indigenous medicine. Its first task was the registration of vaids and hakims and the creation of a roster of vaids and hakims able to produce government certificates, which could be accepted when granting leave and wages. The Board consisted of prominent but lower-level Indian Civil Service (ICS) members, as well as several vaids and hakims who occupied other positions of importance in government and society. Decisions regarding the various practitioners who approached it for recognition were made on an *ad hoc* basis, which took into account its' member's perception of the reputation of the practitioner or institution in question.

The Medical Board in the pre-1919 period had primarily existed as a forum for collection of material – the deployment of decisions – about medical planning in the different provinces of British India. It was primarily occupied with implementing bureaucratic measures and dealing with specific, non-critical issues that arose. The central medical department acted as an intermediary between other imperial/international entities and the provinces, gathering and disseminating information regarding plague, famines and health epidemics as it pertained to bodies that migrated, and also collating data and information relating to hospitals, populations, medical innovation and technologies and medical registration in the different provinces. However, while the GOI maintained its disengaged stance on indigenous medicine, the medical councils of the Provinces had more pressing concerns, including financing their newly transformed administration.

An early test of the local government's willingness to adjust to the new administrative circumstances came in 1923 during a discussion of pensions and salaries of the new cadres of middle-ranking Indian officials. The expansion of the administrative staff was of course a necessary corollary of the increased purview of provincial governments, and they had to follow the protocol already in place for receiving payment. A memo had surfaced in early 1923 enquiring about the granting of medical certificates that were then submitted to government as proof of illness, and generally used in an assessment of said patient's right to take a leave of absence from work. Although the Indian Medical Registration Act forbade non-registered practitioners from certifying illness for documentation used by the colonial state, it is clear from this memo that in certain cases, and particularly when a 'Vaid (was) known to the government', exceptions were made.[15] The purpose of this discussion was, therefore, to identify those who could and could not grant certificates, and for whom. Medical leave was available to government staff, and district staff could be granted leave on analogous terms to those that applied 'in the case of government servants of a similar class.'[16]

This was taken to mean that Boards were barred from accepting certificates from vaids or hakims. The onus was therefore upon the applicant to obtain a certificate from a government-registered doctor, so as to replicate the conditions of application of a government servant. However, in a newer draft of the legislation that was being considered by the provincial medical council in the early 1920s, the wording had been changed: instead of insisting on a similarity to government classes with regard to the application, it merely stated that leave could be granted up to the time allowed for government servants.[17] This left a loophole in the wording that provided for the inclusion of a certificate made by a variety of practitioners. The view of the IMS was that this was an omission that needed to be amended to reflect the regulatory effect of its predecessor: 'The (medical) Board has or can exercise some control over medical men professing the Western branch of medicine ... there is no similar control over Vaids or Hakims,' argued the IMS.[18]

The rest of the committee was split on the issue. Most were in agreement that without a system of registration for vaids and hakims, it was clear that there was a scope for quackery within the system. Interestingly, this was represented as both intentional and non-intentional: while the potential for intended abuse was noted by those wanting to fake illness, it was the Ayurvedic and Unani systems that were considered to be fundamentally unscientific and therefore untrustworthy; it was the methodology, and not the practitioner, that was considered faulty in this instance.[19] On the other hand, it was also proposed that the potential for harm was limited, especially if the leave being applied for was short and the level of employment of the patient was junior.[20] This sheds light on the underlying dynamics of the issue: ultimately it would be the body in question and its status in society that would determine the diagnosis and treatment it would receive; if the body was Indian and the individual was sufficiently low-ranking, the claim to intellectual authority in understanding its functioning could be made by a practitioner the state deemed to be outdated and unscientific.

However, the lack of consensus throughout the province as to the recognition of expertise was one of the challenges the committee had to face, and it makes evident the extent to which practice varied from region to region; the idea of a 'provincial' take on medicine was one that held little resonance when it came to extra-state practices – for instance, when these suggestions were sent to the advisory board.[21] At the meeting, the advisory board found that district boards of Allahabad and Gorakhpur were accepting these certificates. However, other regional boards conformed to the Indian Medical Services Act of 1916, which did not allow for these certificates. A compromise came in the form of a suggestion from the Sitapur district board, the delegate of which called for a committee of seven vaids to monitor, regulate and ultimately register practicing vaids, and an identical though separate committee for hakims to do the same.[22] It was this direction that the provincial government decided was the right and, in fact, necessary one in determining a solution to the problem of medical certificates. At the same time, it could provide some stability throughout the region, which would hopefully ground ideas of regional unity in relevant and innovative new policy.

The idea of regulation also took on a more metaphorical meaning within the context of the debate, even following the redirecting of attention away from certificates and on to the registration question. In later commentary following the resolution, members of the provincial Medical Board – and Ivo Elliottt, IMS surgeon general for the UP in particular – expressed alarm at the variance amongst district boards with regard to these important matters of policy. Mr Elliott stressed the importance of creating a strong provincial board that didn't pander to the districts and that maintained the authority of the Indian Medical Board that existed at the national level. This makes evident the gaps in the presumed dominance of Western/allopathic medicine, and the flaws in late imperial bureaucratic infrastructure: despite the precisely worded acts of the 1910s that forbade the official recognition of any documentation produced by a non-registered/non-allopathic practitioner, for those on the margins of power – in this case, the Indian officers and servants of low rank – the indigenous systems were still widely employed and were in fact being adapted to meet the superficial requirements of imperial functioning. The certificate in this instance is the marker of progress, and the decision of the Allahabadi and Gorakhpuri district boards to accept the certificate of the vaid or hakim makes evident the modernization of the practice of indigenous medicine to fit the new requirements of government. In this transaction, the tension between conflicting notions of Indian and Greater British 'modernity' through science becomes clear, as do the limits of both systems.

The contours of the debate changed swiftly to the idea of registration of vaids and hakims, a question that would occupy the provincial Medical Board for the rest of the decade. In 1923, the Board developed three categories of registration:

> Those who pass from the state-aided Ayurvedic and Unani Colleges will be allowed to issue certificates of leave to persons drawing salaries up to Rs. 250 p.m. and to grant age certificates. Students of state aided Ayurvedic and Unani schools and institutions recognized by the Board of Indian Medicine will be allowed to grant leave certificates to persons drawing salaries upto Rs. 100 per month. Persons holding certificates from well-known Vaids and Hakims or of any institutions of repute or persons approved by the Board as possessing sufficient professional knowledge will be allowed to grant leave certificates to persons drawing salaries up to Rs. 50 p.m.[23]

These regulations were intended to apply to all servants of government. However, gazetted officers would still have to see surgeons certified by the IMS in order to gain certificates, and non-gazetted officers would have to have their certificates co-signed by IMS representatives. Interestingly, it was the board, and not the committee of Ayurvedic and Unani practitioners which had been established for the purpose of sorting out registration regulations, which set the tasks that registered practitioners could now perform. The IMS representatives noted that the committee of practitioners had set the pay-scale limit from Rs 20 to Rs 150 lower than the board, and made no mention of creating certificates for high-ranking officers. The IMS also had trouble with these issues. Furthermore, the IMS was concerned with the use of poisonous indigenous drugs by practitioners, in accordance with the United Provinces Medical Act of 1917 that granted the same rights to allopathic practitioners. The absence of a necessary license permitting the use of drugs and remedies in the pre-existing medical codes created to regulate Western medicine resulted in a lack of precedent for making this a necessary precondition for indigenous practitioners.

The Indian Medical Services also had some concerns about the granting of these certificates to doctors popular in the area. It was eventually determined that only those practicing in the United Provinces would be allowed to take advantage of these provisions, which prevented visiting practitioners from practicing when in the UP. Furthermore, the registration would only be available to the following categories of people:

A. Vaids or hakims who hold a degree or certificate of the proposed Ayurvedic and Unani colleges or schools or a degree of Indian Medicine of any university established by law in India.
B. Vaids or hakims who have passed an examination held by the Board of Indian Medicine (BIM).
C. Vaids or hakims who have passed an examination held in the province or outside, which may, for this purpose, be recognised by the BIM.
D. Vaids or hakims who possess a certificate from any well-known vaid or hakim of the UP or outside of an institution recognized by the BIM and who have practised their profession for not less than five years.
E. Vaids or hakims who in the opinion of the BIM, possess sufficient skill and knowledge of their profession and who have practiced their profession for not less than five years.[24]

The fees that vaids and hakims would receive for the certification of age and illness were set at Rs 10/Class A practitioner, Rs 5/Class B or C practitioner. The IMS also suggested that those registering without the benefit of training should only be registered for an initial year, in which they could take measures to gain some instruction from an institution or a college that would allow them to become Class A practitioners. Those who fell within the Class A category would be eligible to vote for an election to the Board of Indian Medicine, would have the power to issue certificates of leave and age and would be eligible for employment by local bodies. With reference to the Medical Board's concern regarding poisonous drugs, it was decided that these practitioners would have the right to stock indigenous poisonous drugs included in the list published by the BIM and would be subject to an 'observance of the conditions laid down for the allopathic practitioners under the United Provinces Medical Act of 1917'.[25] The measured yet steady increase of the rights and responsibilities of these practitioners reflects the slow progression of the provincial government in both indigenising aspects of governance and absorbing Indian actors into the lower levels of the colonial administration. The regulation of indigenous medical practitioners was the first step towards their initiation into the functioning of the revamped colonial state as 'professionals'.

## Resistance and regionalism: From registration to certification

Despite the attempt to introduce legislation necessary to implement the system of registration, and despite the unanimous support of the varied committee members, Ivo Elliott, director of the IMS in the UP, severely cautioned the government against its adoption. In a letter responding to the proposal, Elliott argued that all was not as it seemed with regard to the implementation of the government policy on indigenous medicine that had emerged in the 1910s. Elliott's main concern was with regard to standards of education and learning. Elliott conceded that, when taught in an institution approved by government, Ayurveda and Unani medicine were 'taught according to scientific principles'. However, the BIM's system of registration would incorporate those who had been educated outside of these institutions, for whose knowledge the government could make no account. Furthermore, Elliott mentioned some problems with the institutions already in existence: despite grants by government to set up institutions, not much of the money had been spent, and the two institutions that had received aide in the UP were barely running.[26] Also, the concern about poisonous drugs was stressed by Elliott, who feared for the intentional and non-intentional substance abuse that would inevitably occur if unskilled practitioners were allowed to use chemicals whose effects they did not understand.[27]

In a surprising move, the BIM turned down Elliott's request to put the registration issue on hold. While recognizing the importance of his concerns, the BIM argued that his resistance was 'too rigid' and that the issue was more urgent than Elliott had anticipated.[28] J.M. McClay of the IMS wrote on behalf of Elliott to the Board, mentioning these critiques, and also bringing up the issue of exams and registrations that had been left unclear in the draft of the registration guide. The Board, however, responded that, despite the concerns brought to light by the IMS, the district boards should be allowed to make their own decisions about whether they should adopt a registration process, and that if they chose to adopt one, they should send in a detailed outline of how they would determine medical proficiency. This was a double blow for the IMS in the province, both in its rejection of the IMS' plea to put off the registration issue until the institutions were better equipped to provide education and hold examinations and also because of the decentralization of decision-making from the provincial board back to the district boards. In an interesting move, the BIM argued that it wanted to uphold the proposal because of the difficulty that low-ranking and poorly employed servants had in obtaining certificates, a concern that the IMS had never addressed. Furthermore, the BIM decided that indigenous practitioners engaged by the Board could provide a certificate for anyone whose pay was less than Rs 30, and who would not receive more than one month's leave, regardless of said practitioner's level of training.[29] In the Board's own words it '[did] not deem the registration of Hakims and Vaids to be necessary'.[30] Granted, a practitioner engaged by a district board would have had some training and would probably be the product of an institution – a necessary precondition set out in a letter to the Lucknow district board, which stated that practitioners from the Vaid Sammelan in Cawnpore, the Tikmilul Tibb in Lucknow or the Tibbia Academy in Delhi were the institutions recognized by the state as producing practitioners who could certify illness/age without counter-signature.[31]

The district boards, however, did not agree to the terms. The first opposition came from Rae Bareilly, in Lucknow district. While the board agreed to accept certificates, they argued that a committee of seven experienced vaids and hakims be established, as previously discussed, to register practitioners who had not been educated in the state-aided institutions. Furthermore, they insisted that an examination be set and that vaids and hakims must pass it. They also suggested that an amount of Rs 50 be allotted to each applicant who wished to be registered – and that this money come from the government, as the board could not afford the expense.[32] In Meerut, the question of fees arose as well, and it was decided that a vaid or hakim would have to pay Rs 5 to the

Board in order to register. They also decided that only literate vaids or hakims could register. Overall, however, most districts complied with the standards set by the BIM, and by the beginning of 1926, medical certificates prescribed by vaids and hakims were widely accepted. This led to a default system of registration of vaids and hakims, setting the precedent for later legislation: in order to write a certificate, one had to have gone through the process of notifying the board of one's compe- tence as a practitioner, which involved proving one's capabilities. The lists of practitioners engaged by boards thus became the first complete and widely used lists of capable practitioners in the United Provinces. The question of a formal registration policy found its way, if somewhat circuitously, on to the agenda of the district boards, and procedures were being put into place to register hakims and vaids in the various districts of the United Provinces.

This shifting allegiances and opinions of the various factions involved in decision-making make evident the state of upheaval – and the result- ing room for flexibility – ushered in by the advent of a dyarchic devolution of power. The IMS representative Sir Ivo Elliott insisted on a tight, centralized policy that would result in a uniform schema that reflected an abstracted ideal more befitted to the imperial centre; his concerns and suggestions required an expenditure that outweighed the ideological interest of the public health project in post-war India. The members of the Board of Indian Medicine were reticent to let the IMS determine policy for the region, and exercised their power as a new insti- tution that reflected the political moment: by being in line with the redistribution of colonial power and rejecting earlier colonial catego- rizations of knowledge and value that reflected a now outdated imperial project, their approach reflected the financial and political organization of post-war UP. However, the district board's plea for higher standards of care, fuelled by a more localized experience of the implementation of the new system, raised the beacon of the value of care as a major concern, placing itself within the ideological centre of the two other opinions, but also completely apart from the urge to balance new poli- tics with older ones. The push for the registration of practitioners locally made evident the extent to which the bureaucratic changes of the post- dyarchic era were felt at the most local levels of governance. Finally, the triumph of these concerns about quality as the driving force behind fur- ther medical reform, as we will see in the next section, makes evident the localization of political decision-making during this period.

The finalization of the registration policy reached occurred when in 1925 it was decided that the United Provinces Medical Act would be amended to incorporate practitioners of the indigenous medical systems into the ranks of those registered according to the criteria set out by the Medical Council through the Board of Indian Medicine.[33] The majority of the officials involved in the decision, including the Director of Civil Hospitals and the Minister of Education, posed no objection to the passing of this sort of legislation, and in fact spoke positively of the accrued profit, which would result from collection of registration fees, for the state. The registration of indigenous practitioners was in fact recast as a subject of more general importance as the relevant officials schemed to promote registration amongst Indian allopathic medical practitioners. G.B.F. Muir, the secretary of the Medical Council, made the case that the legislation in the UP mirrored that entrenched in England, with the UP Medical Council mirroring the General Medical Council London. In both cases, registration was on a voluntary basis; however, as Muir pointed out, the General Medical Council held huge authority in the professional life of doctors. 'Any physician who fails to register himself, or who is struck off from the register, *ipso facto* becomes subject to a strict boycott, the public at large regarding him either as a quack or as a scamp, as the case may be, and no reputable doctor will associate himself with him professionally in any way.'[34]

The Board of Indian Medicine's importance was redefined when in November 1934 official evaluation was introduced, permitting the certification of vaids and hakims practicing outside of the scope of government. Four certificates were established, demarcating the two indigenous medical traditions (Ayurveda and Unani) and the two categories of institutions offering training (school or college).[35] The differentiating factor between a college or a school, as delineated by these certificates, was training in surgery. This distinction is of significance, as the most lasting of the colonial critiques of both Ayurveda and Unani had always been the absence of a refined surgical method. Whereas indigenous understandings of chemistry and physiognomy were not considered capable of possessing the same validity as the Western biomedical reckoning of these subjects, colonial scholars acknowledged that a chemical basis for pharmacology and a theory of anatomy underwriting physical understandings of the body were employed in Ayurveda.[36] Some doctors were aware of Susruta's rhinoplasty; nonetheless, surgery was generally thought by colonial authorities to have been a sloppy addition to indigenous practice, based on inaccurate observations of biomedicine in Western hospitals.

Questions of education and standards of practice were easily resolved by the Board, due in part to preceding discussions about these topics.

The registration of practitioners had set the stage for their eventual certification, a move that was represented by the Board as being the natural outcome of these policy moves. Of great and most probably unexpected concern, however, was the issue surrounding the indication of a practitioner's level of certification. It was at first decided that it should be indicated after the name of the practitioner, though the exact wording was debated amongst the ministers. It was argued that if the letters L.A.M., which stood for Licentiate in Ayurvedic Medicine, followed the name of the doctor, patients might be deceived by the resemblance to M.B.B.S., the letters used by those trained in allopathic medicine. Instead, it was suggested that vaids and hakims add an additional indication that the Board of Indian Medicine had certified the Licentiate – for example, one Vaid Sharma, who had received a Licentiate in Ayurvedic medicine would now be able to sign his name as Vd. Sharma L.A.M.B.I.M, or Vaid Sharma L.A.M. (Bd. of Indian Med.).[37] Eventually, the letters were dropped completely, in part not only for fear that the use of initials by homeopathic and American doctors was already sufficiently confusing and deceptive but also because it was felt that the 'old Sanskrit and Arabic titles are most suitable for vaids and hakims than Western letters even though the instruction is now given more or less on Western lines'.[38] The board eventually went as far as including a clause on the Licentiate certificate, barring those in possession of said certificate from using Western-style initials.

This makes evident the anxiety felt by the provincial government about the modernizing project underway. Such fear of 'deception' prompted H. C. Buckley, Public Health commissioner of UP, to state that 'under no circumstances should letters be allowed to appear after a name'.[39] Even after it was agreed that initials would be banned, Buckley was still concerned that the terminology was deceptive: he felt that the term Licentiate was associated with a stage of education in Western medicine, and that it would be 'undesirable and confusing for the public to use it in connection with Ayurvedic or Unani Medicine'.[40] While Buckley's conservative attitude reflected his role within a bureau that had remained more central than provincial, his representatives tended to take a more cautious approach to the granting of authority or legitimacy to native practitioners. As Ivo Elliott's objections to registration had made evident nearly a decade earlier, the regulation and modernization of these systems was an economic and political necessity following the reorganization of the provincial governments, as the new government employees would require treatment that conformed to the official protocol. This reticence was echoed more generally, especially when the devolution of power would result in a devolution of authority to those actors who had only recently been systematically marginalized from systems of power.

At the same time, the Board of Indian Medicine was a regional institution. The colonial logic behind the modernization of the traditional knowledge systems, and indigenous medicine in particular, opposed the transfer of official recognition and licensing from one province to another. The Board of Indian Medicine was a product of the politic of the United Provinces. Its pronouncements had little cultural or social currency outside of the province. Moreover, the Board had no legal authority outside of the UP, nor did UP recognize similar degree-granting bodies in other provinces or presidencies. Vaids and hakims had no legal bases for practising outside of the region in which their degree had been granted. The privileging of provincial authority in decision-making had effectively halted the free flow of these services by regulating their practice under region-delimited law.

It was upon both of these grounds that the Board of Indian Medicine contested government edict. K.P. Srivastava, chairman of the BIM, wrote to the medical department, stating that the degree programme that vaids and hakims receiving an L.A/U.M.S received was rigorous and long, and that the fees and effort expended by the trainees were considerable. At the same time, the growing difference between a licensed practitioner of indigenous medicine and a practitioner who had not followed a state-approved training course was by this point significant. Srivastava argued that without the 'stimulation' of a Licentiate at the end, the vaids and hakims would question as if their effort was unwarranted. Though it was illegal to call oneself a holder of an M.B.B.S. if one had not done the course, calling oneself a vaid or hakim required no 'formal' training at all. Invoking the regional specificity insisted upon by the Government, the BIM pointed to Madras presidency, where the term Licentiate had been maintained, and had, according to the BIM, gained favourable recognition throughout the subcontinent.[41]

The purpose of the BIM, according to S.P. Shah, secretary of the local self-government department in UP, in a statement circulated to finance ministers in Sindh, Burma and Bengal, was:

> To advise Government on all matters connected with the organization and development of the Ayurvedic and Unani systems of medicine, [to] prescribe courses of study for examinations, maintain a list a of registered Vaids and Hakims, and to distribute funds placed at its disposal by the Government to such Ayurvedic and

Unani dispensaries established in urban and rural areas and make a free distribution of medicines among indigent patients and educational institutions affiliated to it where the indigenous systems of medicine are taught; the amount of the Board's grants varying with the popularity and utility of each dispensary or institution.[42]

However, the line drawn between the Government and the BIM, as well as the BIM and the local practitioners, was unclear at best. The decisions of government on nomenclature, and in particular anxiety concerning the 'deception' of the public, had little to do with the actual perceptions of the medical choices faced by the masses. Moreover, the primary purpose of official certification of vaids and hakims had not been to inform the public but to inform the government itself. The certification of indigenous practitioners had arisen as an issue in medical cases where the individuals in question were employed in lowest branches of Government (as denoted by salary level and leave allowance). The need for expanded health coverage for government employees had been a defining factor in the decision to regulate Indigenous Medicine. Regulation had not been instituted in order to protect or inform those patients outside of government service.

A complex cultural coding system had to be developed by vaids practicing among the general population, in order to assert legitimacy and authority. The notion of 'deceit', which preoccupied the Board, reiterated a differentiation between Western allopathy and indigenous medicine, in which the former was understood as scientifically superior to the latter, and so was of limited interest to vaids' patients. Homeopathic, American and quack practitioners could 'deceive' their patients by claiming allopathic credentials – and charging allopathic rates – as the medicine they used was not as familiar to most Indian health consumers, who relied on Ayurvedic and Unani medicine and practitioners. However, while vaids and hakims could possibly dupe their patients into paying allopathic prices for Indigenous medical techniques, the alternative provided by government did not leave them an option for the appropriate pricing that reflected their long training period. Buckley's paternalist concern mostly reflected his refusal to fully acknowledge the scientific authority of non-allopathic medicine. Attempting to be conciliatory, Buckley agreed to a demarcation: though he insisted that he could not be led astray by the 'bad example' set by the Madras government, he changed the title to read 'Certificate in Ayurvedic/Unani Medicine', therein giving the vaids their due.

The Board of Indian medicine remained an important yet rather ineffectual institution of the state, acting as the liaison between the government and practitioners, yet lacking the ability to assert much power or authority over either. While capable of making recommendations to government, it lacked the power to enforce change, as witnessed by the inability of Srivastava and his board to advocate for a recognizable demarcation of profession. At the same time, as we will see in the next section, while it had the power to assess indigenous medical institutions and report on them to the government, it lacked the power to exert any control over their functioning. Finally, despite the rhetoric devoted to the upkeep of the BIM and its work, the advent of the war resulted in the continual reductions of its funding, which resulted in staff cuts. By the early 1940s, it could no longer afford to keep vaids or hakims not already employed by other facets of government, therein lacking the perspective of community-based representatives.

\*    \*    \*    \*

The significance of the reform of the Indigenous Medical Systems is best understood in terms of the political processes that it enabled. As a non-contentious issue transferred to the provinces as the minor part of a major political revamp, the Indigenous Medical Systems were a *tabula rasa* upon which the newly elected Indian officials could impose their shifting notions of good governance. In the early dyarchic period, the concerns of the government were local and regional, and the rhetoric employed and policy devised reflected the concerns of a government delimiting a political space. The reforms envisioned cautiously approached the subject, focusing on the professionalization of the practitioner as the key to the modernization of the systems. In this period, the provincial government focused on the needs of the government to fuel the overhaul of the system.

At the same time, these reforms have remained notorious because of the strident ways in which they came to frame subjects of governance that had been previously considered ungovernable. The haphazard way in which they came together through impromptu surveys, debates about structure, discussions about standards and rigor and insufficient data about the daily practices of the systems can paint a picture of policy forged out of scramble. This representation belies the creative and inventive ways in which the newly autonomous governments of the provinces worked to reshape and, frankly, to reinvent the spectrum of governance in the provinces. Free from the watchful eye of the imperial state (which had abdicated both responsibility and interest in this

variety of reform), and not yet tied to the grandstanding and opera-
tionalizing forces of formal party politics, the 1920s and 1930s present
the historian with the opportunity to witness a unique moment within
the history of subcontinental governance. The remarkable achievement
of these reforms is also witnessed, as we shall see in the following
chapters, by the way in which their structure and logics lasted well into
the era of Indian independence.

The post-1937 era of Congress ushers in an end to this kind of
autonomous creativity, while the expansion of the Indigenous Systems
of Medicine and their incorporation into early discourses of political
planning were part of a more substantial imagining of governance in
an independent state. In this period, Indian officials began to explore
the macroscopic nature of the nation, framing UP as a microcosm of an
emergent nation, and framing policy that reflected these political ideals. The political framing of health reform as a national project rid it of
its discrete regional focus and adopted the rhetoric and ideology of an
all-India project, recasting reform as the first step towards unity. In so
doing, the different factions began to emerge as an important factor in
the framing of both discourse and praxis: the importance of community, tradition, region and locality began to emerge as key determinants
of every aspect of the political process. Where the Indigenous Medical
Systems had sufficed as a category of analysis and a subject of reform in
1921, from 1938 the particular contexts of Ayurveda and Unani started
to dominate policy reform.

# 5
# Planning Through Development: Institutions, Population and the Limits of Belonging

In this chapter, we explore the Congress government's attempts to move beyond the scope of the professionalization of vaids and to engage with a more meaningful revamping of the health infrastructure of the United Provinces. In the previous chapter, the rise of the practitioner as the arbiter of 'traditional' systems was made evident through the unravelling of health policy based mostly on his or her credentials as a subject of 'modern' medical professionalism. Ayurveda was introduced to the government through the traditional medical practitioners consulted by their lowest-ranking servants: the right to claim sick-leave had also resulted in the necessity of a doctor's consultation; vaids were of use to the government in so far as they could participate in its bureaucracy by providing this service to its newest employees. Government policy through the late 1930s had focused on the certification and registration and regulation of vaids, therefore cementing them as the mode through which the state encountered Ayurveda.

However, the ascension of the Congress to an official position of power in the provincial legislature of the post-1937 period introduced a variety of shifts in governmental practice. The Congress government saw in Ayurveda a potential mode for a furthering of national unity based upon the consensus that the 'authentic' scientific systems of the subcontinent were something worth investing in. Despite the concerns of Muslim members who questioned the subtle exclusion of their own traditions in the catch-all rubric of an 'authentic' Indian identity, the Indigenous Systems of Medicine, articulated mostly through the experiences of Ayurvedic practitioners, became a concern for the government beyond the immediate demands of the practitioner's rights.

In this chapter, I explore the next phase of the story for the Congress: namely, its investigation of and investment in the spread of the

Indigenous medical systems through the creation of new institutions and the consolidation of those already in existence. This began with a survey of the pharmacies and dispensaries already in existence in the late 1930s, which resulted in the mapping out of a demographic survey of the United Provinces via regional access to health services. Conceptualized and justified by the Congress government as the means through which services would be developed and deployed throughout the province, it set the stage for more profound discussions about the boundaries of the urban, the rural and those areas completely marginalized from medical services connected to the state. This process of data collection and the ordering of information harkens back to an earlier experience of colonial biopower, reframed in this instance as the 'development' of society. At the same time, where the political vacuum of the 1920s and early 1930s had created the space for creativity in policy-making and political imaginings, the dominance of the Congress also introduced a more rigid schematic in planning and implementation, ushering in a more formalized era of late colonial biopolitics.

However, the dream of secularist values embedded in discourses of indigeneity, which Benjamin Zachariah has identified as the Congress' language of national legitimation, was not necessarily the obvious outcome of these endeavours.[1] As we saw in the last chapter, the concerns of the Congress Muslims had been legitimate: Ayurveda was becoming a catchall category for anything medical and indigenous; furthermore, Ayurveda was a system highly coded as Hindu. The latter part of the chapter explores the construction of a different sort of nationalism, reliant on a different discourse of indigeneity through the example of the Rishkul Ayurvedic College of Haridwar. This college used the new government's interest in Ayurvedic institution-building to establish legitimacy for communalist organizations to which it had ties, resulting in a crackdown against it once the RSS was seen distributing materials on its premises and through its members.

Taken together, these experiences of institution-building raise interesting questions about the state of the colonial institution in this period. To take up Marks' question one final time, what is colonial about late-colonial institution building? Was colonial rule something that institution-builders laboured under, or was it a logic they happily adopted? Furthermore, how was the end of colonialism imagined in the planning of these institutions? Developed in the early 1940s, these institutions emerged on the cusp of a global upheaval with very local resonances, and when the independence was becoming a steady reality more than a cause to fight for. They also emerged at a moment when the central mechanism of public health planning in Delhi was taking up the challenge of transnational health organizations, and doling out more and more responsibility to the provincial governments.[2] Indigenous medicine, therefore, in this period, was deployed not only as an ancient cure for modern ills but as a traditional vehicle through which the nature of independence could be determined.

## Creating 'standards' of care: Reforming the BIM in the Congress era

The dominant focus of the Board of Indian Medicine (BIM) was clearly the increased professionalization of the medical services so that they could better serve the needs of the state. However, implicated in the notion of professional improvement and bureaucratic efficiency was a commentary on the larger state of the system under consideration. While it was clear that the thrust behind the drive for professionalization was the thrifty efficacy of the provincial government, the rhetoric used to justify the changes often involved a broader critique – as well as a nuanced, if vague, plan – to uplift the systems of indigenous medicine more generally, in an effort at making tradition applicable in a modern era. In the early dyarchy era, the changes envisioned were cautious ones as Indian politicians navigated an unprecedented political system in which they had very little power to implement their decisions.

In order to gauge the extent to which the Congress differed from the earlier provincial government, it is necessary to comment briefly upon its development in the UP. Gyan Pandey has traced the evolution of the Congress in UP in the interwar period, arguing that Congress grew to power through a combination of Seva Samiti aid and local fundraising, and made itself meaningful through the creation of district and *tehsil* Congress committees which would keep vigilant watch on local political developments.[3] Support for Congress peaked in 1921, and ebbed in 1927, when public opinion split between support for Nehru and Malaviya. In 1934, it was decided by the All-India Congress Committee that Congress should contest the elections to the legislative assemblies, and public support for Congress grew and grew, resulting in remarkable gains for the party. This was partly due to the wider context of imperial politics, as well the Simon Commission and the protests resulting from it, along with the Irwin Pact, made their mark on the public imagination.[4] The Government of India Act of 1935 expanded the suffrage exponentially and abolished dyarchy, transferring responsibility for all portfolios to the provincial governments, where they would be looked after by ministers supported by the provincial legislatures, and who would report back to Delhi. In 1937, the Congress successfully contested elections throughout the UP and became the party of government in the UP. Though they would resign in 1939, several key changes that affected the indigenous medical systems were made before the Quit India movement.

As we shall see, the election of a Congress government in the provinces did by no means result in the democratization of politics promised in the 1935 Act in a way meaningful to the population. As David Washbrook has argued, the lack of both resources and organization at the provincial level resulted in very messy local politics concerning fights over local patronage and sharpening communal tension.[5] Gandhi's order to 'resign the ministries' in 1939 ostensibly saved Congress from itself by reasserting anti-colonial resistance as the focus of politics. Francis Robinson and Paul Brass also both argue that the communal tension, characterized by the reification of a Muslim minority identity first by the Raj and then by Congress, was a key component of UP politics, though they are in disagreement about the use of the Islamic minority identity amongst Indian Muslims. This had very real effects when even the most uncontentious issues of policy were being constructed.[6] Social difference had to conform to the larger goal of anti-colonial nationalism, especially concerning communal issues. However, the Congress government did invest in certain social issues beyond the Gandhian focus on Harijan uplift. As we shall see, Congress diverted attention away from communal issues by focusing on the uplift of Indians along class and mobility lines, an edging away from the issue of minority rights.

With the possibility of war on the horizon and the continued devolution of British interest in all aspects of Indian political life, the arm of the Raj ceased to stretch into provincial politics as far as it had earlier on, especially with regard to institutions that were not politically contentious.[7] Congress politicians had, for the first time, a political office, backed by the Indian masses, from which to flex their political muscles. While research into the police and the military reveals that the centre still maintained firm control over provincial politics, an exploration of the social services provides an alternative view.[8] Medical reform in the Congress-led era makes evident the extent to which Indian politicians took steps to overhaul the system, beginning by increasing the power of the BIM and transforming it into a properly statutory body, backed by legislative power. Moreover, an exploration of the rhetoric and ideologies that fuelled these discussions reveals that

Indian politicians used social issues, like the reform of Indian medicine to transcend the realm of provincial politics, and to begin to imagine models of governance in an independent state.

The initial discussion of registration and certifications, centred as they were around the practice and the authority of the practitioner, evoked a larger commentary on the state of the system that centred upon the level of achievement of the practitioner. The earliest government documents concerning these changes refer to the 'sense of public responsibility' of those practitioners whose trusted word was being evaluated.[9] Ivo Elliott, aforementioned as a critic of both dyarchy and the indigenous systems, used this notion of civic identity as the means to strengthen his dominant argument for a strong centre and against the devolution of power. However, placed within the context of dyarchic notions of civic identity, the conception of indigenous practitioners as unconcerned with public responsibility was significant on two fronts: firstly, as an attack on the ability of practitioners to serve as professional medical providers; and secondly, and more importantly, as an attack on the ability of those still caught up in the 'traditions' of South Asia to serve their society as modern citizens.

While the consequent actions taken to register and certify practitioners were formally intended to bring indigenous practitioners into line as medical professionals, the simultaneous result of this move was their induction into the realm of those accountable for civic life through responsible governance. The members of the medical board envisioned this transformation as the litmus test, casting the move towards professionalization as a social experiment that would gauge how those prone to conceive of the world in culturally and ethnically bound terms would deal in more 'global' terms. The original end goal of the government was the expansion of this particular service, namely, the granting of medical certificates, to the majority of the population. However, in its preliminary stage, official government servants were barred from participation – only employees of local bodies could partake of these services. 'Until [the medical board] has some experience of how [indigenous medical practitioners] exercise the power of granting certificates to employees of local bodies,' stated the report of the Medical Board, 'there should be no question of empowering them to grant certificates to Government servants.'[10] This relegated the new service to Indian bodies: having traditionally been understood as fit for the potentially dubious 'science' of the traditional medical systems, they were now considered the appropriate subjects for socio-political experimentation. The potential for exploitation lost the capacity for risk when practiced upon them.

The initial concern of the Medical Council, and the subsequent subject of improvement, was thus the practice and ethics of the practitioner. The first decade of policy brought the traditional practitioner into line as a proponent of modern medical care using various techniques developed in the previous decades to regulate and legislate colonial science and medicine. The expansion of the subject group for this variety of political disciplining to include indigenous medical practitioners previously excluded from colonial machinations of medical governmentality broadened the general discussion about the state, even of the allopathic field of medicine and its practitioners of Indian descent. The registration of indigenous practitioners, as discussed earlier, called into question the issue of registration more generally and the failure of the United Provinces Medical Council to assert its dominance as the arbiter of practice and care.

In the United Kingdom, the General Medical Council had come to dominate the public view of the medical profession; in India, the equivalent Medical Councils still remained largely irrelevant organizations. It was unsurprising to the UP Medical Council that indigenous medical practitioners or those who consumed their products and services would have been unaware of the Council or its attempt to adjudicate the practice of medicine. However, the extent to which those Indians who did achieve medical degrees in the approved Indian institutions of allopathic medicine failed to recognize the authority of the board was troubling to the Council. As Muir noted, 'failure to register with the Council, and even removal from the Council's register, carries with it no real disability as regards earning a livelihood by the profession of medicine'.[11] His suggestion, which was approved by the other members of the council, was to make registration of all allopathic doctors compulsory. Muir suggested that medical degrees should in fact not be granted by the prevailing institution until students bore the proof of registration. The ultimate question, it appeared, was a larger concern about the role of institutions in the formation of civil society. For Muir and the board, the institutions of government served as the ultimate arbiters of power; India and Indians, however, did not hold similar values: 'so long as in this country public opinion is merely apathetic as to whether a medical practitioner is registered or not, or whether he obeys the council or not, I do not see that mere registration will give the council that authority which it seeks'.[12]

The 'improvement' discourse remained focused on the practitioner until the late 1930s. The proposition of a Bill on the Indigenous Systems of Medicine that would both update and formalize the tasks of the Board of Indian Medicine shifted the focus from practitioners to the state of care they provided. The drafters of the Bill, and those who discussed it, used a rhetoric that implied the resolution of the professional question with which earlier legislators had been concerned. The Bill thus instinctively represented the final step in the disciplining of the traditional practitioner into a modern professional subject: that vaids would articulate their desire to change the state of the tradition through legislative protocol and that they would conceptualize the state bureaucracy as the arbiter of meaning and status represented the natural end to the project of professionalization.

The Bill on the Indigenous Systems of Medicine was initially drafted and proposed by Khan Bahadur Maulvi Fasihuddin of Budaun in 1935. The Bill failed to pass into the final stages at the time of the change of the government in 1937, and V.D. Tripathi, M.L.A. attempted to reintroduce the Bill later that year. However, upon learning of Tripathi's intent, the government intervened, claiming that the Bill was 'harmless', and was in line with similar developments in Bombay; G.B. Pant, the Premier of the Province, decided that Tripathi's Bill should be studied carefully and introduced as a Bill proposed by government.[13] Vijaya Laksmhi Pandit, scion of the Nehru dynasty, and then minister of Health for the UP, agreed, informed Tripathi of the change in plan, introduced the Bill herself in 1938 and headed up the Select Committee to evaluate it.

The intent of the Bill on the Indigenous Systems of Medicine was to uplift the state of the profession, primarily through the revival and reform of the Board of Indian Medicine. The Indigenous Medical Systems, claimed Vijaya Lakshmi Pandit, in her foreword to the Bill, 'are sufficiently popular, not only with the poor and ignorant but also with a large number of educated and well-to-do people'.[14] The cost was not prohibitive, and it was possible to cure a large number of sufferers on a small budget. In a preface to the Bill that was sent around to members of the Provincial Legislative Assembly for discussion, Mushtaq Ali Khan claimed that though the Board had made great strides in regulating those practitioners and institutions registered within it, it was not a statutory body, and the useful measures undertaken by it had failed to result in the 'appreciable development of the two systems in the United Provinces'.[15] Quacks continued to be a problem in the province, and vaids and hakims were at a disadvantage when compared to their allopathic practitioners in the province.[16] Educational institutions were few and poorly equipped, and the funds available for *Dawakhanas* and *Aushadalyas*, Unani and Ayurvedic pharmacies and dispensaries, were inadequate.[17] While the Board's initial mandate had conceptualized and implemented a basic infrastructure for the professionalization, expansion and improvement of the indigenous medical systems in the UP, it was time for infrastructure to be further developed.

The improvements listed made great strides in redefining the Indigenous Medical Systems as a central part of state planning. No longer would the BIM be merely regulating the particular services offered by Indigenous Medical practitioners to government peons. The new mandate of the Board of Indian Medicine focused on three main components: firstly, the establishment, maintenance, regulation and development of institutions where the Indigenous Medical Systems were to be taught; secondly, to retain the register of practitioners and to more forcefully encourage practitioners to register and to curb quackery; thirdly, and most importantly, to invest in pharmacies and dispensaries, and to centre public health schemes around these sites and the expertise found within them. While the effects of the latter investment will be explored in the next section, the provision for public health planning through traditional systems marked a significant departure. Mushtaq Ali Khan and others were careful to note the influence of the Congress government in pushing forward these changes. The improvement of the Indigenous Medical Systems in the UP was therefore implicitly associated with the simultaneous ascendancy of the Congress in the provinces following the contestation of the 1937–1938 elections.[18]

It is unsurprising, therefore, that the debates of the Select Committee reflected the larger struggles being waged in the political sphere. The most predominant of these was the discussion of the balance of vaids and hakims on the Board of Indian Medicine. The 1935 bill mirrored the earlier policy of the Board of Indian Medicine, making provisions for representation by community that were in line with the rationale of the government vis-à-vis the award of communal representation in political matters. However, V.D. Tripathi's initial revision of the drafted bill sought to do away with any sort of representation, arguing that vaids outnumbered hakims and that no special recourse should be made to ensure Muslim representation.[19] Where the previous bill had argued that the head of the committee rotate between a vaid and a hakim each term, and that one Hindu and one Muslim representative be taken from the Legislative Assembly, Tripathi and his supporters envisioned an identity-free document, instead proposing that, in keeping with the India Act of 1935, two members of the Assembly and one of the Medical Council, as well as three members from Local Boards should serve on the BIM. This represented a wide range of politically engaged members at various levels of the administration; however, especially following the results of the 1937–1938 elections and the plethora of Congress-affiliated politicians, the members who would now be charged with responsibility for the Board would likely be predominantly Hindu.

This inevitability was the major bone of contention for politicized Muslim members of the Select Committee that evaluated the Bill, and alternate solutions were offered up to rectify the communal balance. Mr Zahiruddin Farukhi instead proposed a system that called for four hakims elected by registered hakims, and four vaids elected by registered vaids, and one representative each from the major Ayurvedic and Unani colleges and organizations, along with the three members of the Assembly and Council. Vijaya Lakshmi Pandit accused him of marginalizing the district boards and of replacing Muslim politics with rural/urban ones. In the end, the committee agreed to quell the issue by meting out equal assistance to both systems, and retaining the membership of those representatives serving on local boards. While the committee agreed to let the Bill go forth without any formal communal balance of traditional practitioners, it amended the number of politicians to include three members of the Legislative Assembly, one of whom would be a Muslim.[20]

This brings to light the variety of political factors at play, which were embedded in the notion of 'improvement': should the system be reformed to correct the communal balance? Should it be reformed to address the needs of those not represented in the urban power politics of the province? In his concluding remarks on the subject of communal allotments, Mr Farukhi conceded that the current government had previously given assistance to both systems, and that this should continue in the post-Congress era and beyond, as 'it did not look nice to make a departure from that practice when we have our national Government'. Vijaya Lakshmi Pandit had baited him into a comment of this sort: in her attempt to quell his protestations concerning Muslim representation, she had commented to him that 'the idea of Government was to give encouragement to all these Indigenous Systems of Medicine'. The promise of true political power, unmediated by the colonial state, which would have been a more tangible fantasy after the ascendancy of the Congress, was successfully affective: the gravitas of governance was employed to quell the factionalism of individuals with the promise of unity.

It was in debates like these that the faultlines of Congress secularism began to emerge. The promise of unity in governance, however, was not a strong-enough deterrent to stamp out the concerns of the minority members of the Select Committee. Attached to the draft of the Bill presented in the Legislative Assembly was a Note of Dissent on behalf of the committee's Muslim members, signed by Zahiruddin Farukhi, Zahur Ahmad and Tahir Hussain. It read as follows:

> While agreeing with the general principles of the Bill, we think that the constitution of the Board of Indian Medicine should be so devised as to represent all shades of opinion, and the Unani and the Vaidic systems should be treated on terms of equality. The science of healing should receive fair and impartial encouragement, as any other line of approach will not be consonant with the object we still have in view. We think that it was on this basis that the previous Government decided to have an equal number of vaids and hakims on the Board. We are afraid that this principle is not reflected in the present composition of the Board. The Registered Practitioners should elect equal number of vaids and hakims, as hitherto: and the vaids and hakims should elect their representatives separately. There is no reason why a vaid should elect a hakim or vice versa.[21]

Ultimately, Vijaya Lakshmi Pandit's notion of a united Indian indigeneity failed to prevail over the concerns of community. The traditions of each community, even once rendered in political terms, were not categorically interchangeable, even within a political system that attempted their reconciliation.

The political significance of the Indigenous Medicine Bill is in the rhetoric and quality of debate amongst those who drafted it, and also makes evident the reconfiguration of regional politics as nationalist ones. For instance, the Bill details the attempts to discipline the Indian advocate into a modern political subject, which had long been the concern of the UP Medical Board: the revamped Bill addressed the failure of the original committee to curtail quackery and effect professionalization. More profoundly, the select committee-led evaluation of the Bill, and the focus within it on the role of the Legislative Assembly, allowed the political reforms enacted through the India Act of 1935 to be implemented meaningfully. At the same time, the quality and structure of the negotiations, especially with regard to the Hindu and Muslim issue, give great insight into the nationalist reckonings of independent political power and their potential to negotiate political difference. As we will see in the next chapter, the debates over the Indigenous Medicine Bill did little to quell the quacks, or effect a sense of allopathic-style professionalism amongst the vaids and hakims: though some registered and others used the Board-sanctioned 'rights' to their advantage, the public sphere and the bazaar provided the real contexts for their professionalization. However, the reform of traditional medicine created a viable forum for the discussion of good governance amongst Indian politicians, especially as it related to communities and citizens conceptualized as being overly embedded in other systems of authority, as vaids had been before the advent of reform.

The Board of Indian Medicine, while an institution of great symbolic significance, fought throughout the colonial period to maintain control over the practice of vaids and hakims in the UP. Its legislative framework, however, was far from conclusive: significant changes were made to its registration policy just three years after the Bill passed, and continued through to 1949. However, the negotiations over its increased importance and significance as the arbiter of medical practice in the UP provides important insight into the imaginings of governance amongst the generation of Indian politicians already envisioning themselves as leaders of an independent state. The nature of their imagining is significant. On the one hand, their notion of indigeneity, and their attempts to distance themselves as elite, cosmopolitan citizens from the traditions of the masses, echoed those of the forbearers in the colonial government. The discussions of how medicine would work, and how those who practiced it should comport themselves became, in this instance, a site for the negotiation of how the new nation would function amongst those who would eventually come to rule it.

## Dispensing health: Ayurvedic pharmacies and dispensaries under consideration

The governmental medical dispensary played a key role in the perception of medicine within the context of the state throughout the nineteenth and early twentieth centuries. The evaluation and treatment of the health of individuals at medical dispensaries and pharmacies were bound to the larger project of consolidating imperial modernity through the collection and application of 'information' about India and Indians. Dispensaries constituted one of the only spaces where Indians in non-urban settings could acquire Western medical attention and therapeutics in a way unrelated to the larger public health concerns of the government. The government in turn used dispensaries and pharmacies as key locations for the gathering of information about the public health. Reports on pharmacies and dispensaries throughout the colonial period were largely composed of data relating to patients, disease, illness and finances.[22] This manipulation of 'evidence' through a categorization of

Indian life constructed by the British served to further divorce Indians from the systems of power in the subcontinent.

This is particularly relevant to dispensary-based data collection, as the premise upon which the patient was surveyed was different than in more general census-related enquiries: the collection of information in this case was inherently linked to the experience of illness and the seeking out of treatment, thus casting the interaction as one of particular urgency. Patients were asked to fill out forms confirming their age, gender, caste, general background and location, along with more nuanced information about how far they had travelled and the conditions in which they and their community resided, as part of the experience of receiving treatment. Medical pharmacies and dispensaries therefore played a significant role in the collection and ordering of data about the population, thus increasing their institutional value outside of the health infrastructure as well as within it.

Pharmacies were also crucial sites for the development of scientific knowledge about drugs and medicine. As we saw in Chapter 3, the manufacture of indigenous drugs had been a government concern from the 1890s, addressed by the Indigenous Drugs Committee and later by the Drugs Manufacturing Committee. Two aspects were considered central to the manufacturing of indigenous drugs: on the one hand, there was the potential for manufacturing drugs used in allopathic medicine in the subcontinent, which became a priority following the medical shortages of the First World War and its aftermath; secondly, there was the question of drugs used in non-allopathic medicine. The reports drawn up by both committees consistently relied on the information provided by the state medical dispensaries and pharmacies about what worked and what did not, and would often send samples of drugs or medicine back to these institutions for further (or, often, initial) testing. The dispensary was the place to send a new pharmaceutical product after it had been processed in the lab, therein connecting the dispensary to networks of scientists, administrators and entrepreneurs who had a vested interest in developing an indigenous drugs industry.

As we shall see, Ayurvedic dispensaries were 'relieved' of the task of collecting or ordering information, or engaging in scientific research. Instead, their use was determined by the service they provided to ease the burden of government through practices of medical certification. As we've seen, the provincial government of the United Provinces had engaged in the regulation, professionalization and legislation of the indigenous medical systems in order to reduce the amount spent on the medical services necessary to the functioning of the colonial bureaucracy. It consistently conceptualized Ayurveda and Unani within the services they could offer to the appropriate subjects of the health bureaucracy – namely, Indians employed in the lower-levels of service. As such, the Provincial Medical Board eschewed conversations about the expansion of the systems in any way, leaving these discussions up to the practitioners and their institutions.

Ayurvedic dispensaries had a rather singular function vis-à-vis the state: they provided the space in which certified vaids could render the services that in turn benefited the government in its most basic manifestation – basically, as a bureaucracy – through the certification of illness in lower-ranking Indian government employees. The Medical Board acknowledged, of course, that certification was not the only activity of the dispensaries, but failed to involve itself in discussions of just what seemed to be happening within them. This perhaps reveals the true state of the Indigenous Systems of Medicine in this period: while practitioners had to conform to the rigid standards put in place to incorporate them into the medical bureaucracy, the larger question of what they did failed to intrigue government. At the same time, the practice of actively keeping them out of the sphere of health governance, enacted in this instance through the refusal to see the dispensary as anything other than a place that *dispensed* services, provides insight into the way in which the category of the indigenous practitioner differed from his allopathic counterparts.

The early provincial government did not seem to find vaids useful outside of the scope of the service they provided, nor could they conceptualize a connection between Ayurvedic and biomedical dispensaries. The Congress government, however, saw the potential of Ayurvedic dispensaries to be more than the sum of their colonial parts. Under the auspices of a Congress rule, the Board of Indian Medicine began to collect evidence and formulate policy on pharmacies in the late 1930s. The information they collected regarding the location of pharmacies allowed for the mapping of health services in the United Provinces, which both relied upon and contested earlier colonial uses of dispensaries. Not all political discussions were as ephemeral; the major policy reform of the period was the expansion of pharmacies and dispensaries, especially in areas where medical care was sparse and of questionable quality.

## Assessing the constitution of the nation

In 1938, the government solicited a list of pharmacies that distributed and manufactured pharmaceutical products used within the indigenous medical systems from the Board of Indian Medicine. The Board stated that the following were institutions which could be trusted: the Ayurvedic Pharmacy, Hindu University, Benaras; Rishikul Ayurvedic College, Pharmacy, Haridwar; Ayurvedic Rasaymshala, Ayurvedic and Unani Tibbi College, Delhi; Ayurvedic Pharmacy, Kashi-Ras-Shala, Gyan Vapi, Benares; Hindustani Dawakhana, Delhi; and the Unani Dawakhana, Allahabad.[23] In the following years, the Seva Samiti Dawakhana in Allahabad and the Mulchand Rastogi Trust in Lucknow were also included.[24] This mapped out an intellectual and economic network of educational institutions, merchants and practitioners that the government considered to be the legitimate portion of the indigenous pharmaceutical industry of the UP. While the government enquiry had been modest in its phrasing, the enquiry's significance was obvious: soon after they were named, each of these institutions wrote to the government enquiring about the possibility of winning a contract to supply the dispensaries that the government intended to set up throughout the provinces.[25]

By the early 1940s, the list of suitable dispensaries had increased to include pharmacies in Pillibhit (at the Nepali border), Cawnpore, Sitapur, Bareilly, Shahjahanpur, Unao, Saharanpur, Meerut and District Muttra, and to exclude the Delhi-based practitioners, institutions and pharmacies, as per the regional boundaries of Ayurvedic certification.[26] Out of the 48 districts of the United Provinces, 12 had approved Ayurvedic dispensaries, though only 3 (Meerut, Allahabad and Lucknow) had Unani dispensaries.[27] However, not every pharmacy that applied was approved. For instance, the Bhartiya Aushadalya in Vrindavan was not included on the list after its first application. In an enquiry later that year, it was found that the vaid who supervised the making of the medicine, though apparently very well respected in Vrindavan, lacked the accreditation now required by government and was not considered capable of supervising the manufacturing of pharmaceuticals.[28] Similarly, the 'Shri Saniwan Depot in Haridwar' was turned down for only selling 'a few patent medicines of a simple sort', and in status and reputation much below the other pharmacies in the district.[29]

The sale of these medicines was obviously a lucrative endeavour, and the BIM decided that an inspectorate was needed to evaluate the pharmacies bidding for contracts. Though there were discussions to identify appropriate vaids and hakims for inclusion, a default inspectorate of deputy medical commissioners in the provinces reported its findings back to the government, bypassing the Board of Indian Medicine. This was justified with reference to the war, and to the 'need for economy'.[30] K.P. Srivastava wrote to the Department of Public Health (DPH), laying out his recommendations for regulating the inspection process. He proposed that a common pharmacopoeia be prepared, that the use of patent medicines be prohibited, that one Ayurvedic and one Unani inspector be appointed to inspect the dispensaries and that firm rates be carefully examined before being included.[31] Srivastava concluded that the issue of a common pharmacopeia was the most pressing, a stance that was backed and appropriated by the director of Mulchand Rastogi Trust in Lucknow and the Chancellor of the Banaras Hindu University (BHU). The Rastogi trust claimed that their medicines were the most pure, citing the works of other approved pharmacies as inferior; similarly, as the largest institution, BHU claimed that its pharmacy was most efficient at producing quality medicine at low costs and in large quantities. The college made the government aware that other firms were buying its products and offered the government an exclusive contract.

Once the method of obtaining pharmaceuticals was organized, the development of the new dispensaries was set to take place. Having mapped out the location of approved institutions and dispensaries, the government attempted to identify four places where medical intervention was most needed, based on population density and the proximity of an allopathic dispensary: These were a vaid for Auras in Hasangunj Tahsil; a hakim for Karadaba in Purwa Tahsil; a vaid for Gauria Kalan in Safipur Tahsil; and a vaid for Pariar in Unao Tahsil.[32] The government intended to send a vaid or hakim to each location to set up a dispensary there, and had the controlling officer in each district agree to take responsibility for one-fourth of the cost, as well as the residence and upkeep of the practitioner.[33]

Pleas came from other tahsils for similar services. In Bahraich, the deputy commissioner enquired about the possibility of setting up a dispensary in the Bakaina region as the malarial climate created illness every year and medical care was desperately needed.[34] In the impoverished village of Gaucha in Garhwal district, one Vaid Harisharan Sharma approached the government for access to funds for a dispensary, claiming that he was already running a dispensary there and giving much medicine away for free. Harisharan Sharma proved an interesting case for the government to consider. He claimed to have received a certificate in 1927, having passed the degree of 'Vaidya Kaviraj', and included with his letter to the DPH a copy of the certificate, which was handwritten in Sanskrit.[35] His certificate did not conform to the standards set out by the BIM in the mid-1920s, making him technically an unqualified vaid. However, the state of public health was dire in Garhwal, and as his clinic was already established, the situation was ideal for the transfer of funds. Despite its statement that 'the board will appoint only such Vaids and Hakims who have passed the medical examination of the Benaras Hindu University and the Aligarh Muslim University and who could do surgery as well', Sharma was added onto the scheme.[36] The situation was similar in neighbouring Guptakashi, and another grant was transferred to the area. In the village of Kedarnath, also in Garhwal Tahsil, the Rawak Sahib of Kedarnath Temple had funded the entire operation, providing the *Aushadalya* space in one of the temple buildings, paying the salary and upkeep of the vaid, subsidizing the purchase of medicines so that they could be given away for free and establishing the practice there as a major temple site.[37]

At the same time, there is evidence of earlier governmental collaborations with indigenous practitioners – collaborations that underwent a re-evaluation in the late 1930s. In 1921, a vaid and hakim had been appointed by the BIM in an attempt to provide free medical aid to the people of the Etah region. M.Ulfat Rai Sahib, a powerful and wealthy Rai in the region, had supported the venture and had subsidized it heavily until his death in 1926, after which the BIM had absorbed the cost of the two practitioners.[38] Following the decision to develop the pharmaceutical infrastructure of the provinces, arrangements such as this needed to be brought into line, and the government agreed to build two dispensaries in the region.[39] Once again, 'unqualified' practitioners took central roles in a project that had originally been developed to wipe out their ilk. Moreover, despite the government's sharp adherence to the regional boundaries of certification, Aushadalyas in Sikandrabad and Dadri in Bulandshahr district as well as Ghaziabad and Philkua in Meerut district were run by the Dharmarth Ayurvedic Aushadalya Prabandhik Committee in Chandi Chowk, Delhi. The Committee had already established the Aushadalyas before the granting scheme had taken effect, and it seemed more cost and energy efficient to bring existing pharmacies under the control of government rather than to attempt to re-staff them or arrange for a new source of pharmaceuticals for them.

These examples make evident the limits of imperial dominance with regard to public health policy in the region, and indicate the ambivalence amongst provincial officials about what line to adopt. The government had insisted on a system of regulation that undermined the existing medical infrastructure. Although at times it was resistant, the BIM echoed the policy of the rest of the government. The collection of information about districts most in need of help, about pharmacies most qualified to provide medication and about the circulation of practitioners throughout the region was driven by imperial priorities for ordering health care. Imperial concerns generally had little relevance to existing infrastructure of indigenous medical care. However, confronted with a system that was both entrenched and effective, the provincial medical board could not but choose to endorse it; the economic advantages of supporting an already-standing building and an already-situated practitioner were not ones the government could afford to ignore.

This also reveals some of the tensions involved in public health planning. Was the driving factor behind public health planning the reordering of the region so as to prevent 'unscientific' medical care, or was it to address immediate health concerns using solutions already at hand? While the former consideration underlay its ideology, the government generally prioritized its interventions in accordance with the latter preoccupation. Despite the effort made by urban vaids and hakims to obtain the certifications necessary to be employed by government, the experience of established vaids made them equally valuable to the Public Health department. This was also determined in part by their rural setting. Working on the principle that some medical care was better than no care, the imperial government's concern over certification fell away in the mofussil. Where the practitioners in question could neither petition the government nor sign their names in English, the question of too many initials following a name was not deemed an important problem.

This is where the importance of the creation of pharmacies to the evolution of politics in the post-1935 period becomes significant. The effect of the mapping of the UP into areas divided by region and by levels of development, and the formation of policy based on these observations, mirrored the dual concerns of the Congress-led provincial government in their attempts to create policy that addressed the nation as a whole. As we saw in the negotiations of the Indigenous Medicine Bill, the conception of a UP-based microcosm of national unity necessarily relied upon the identification of the different parts of the nation and the ways in which they were both in congruence and in contrast with each other. 'All-India' was, in essence, a sum of its parts, and the implementation of policy that attempted to both survey and address the need of a disparate population, as the development of dispensaries did, necessarily brought to light the two major factions at play, namely community and location.

The community-based argument has its roots in the earliest Congress negotiations of secular unity even if rooted in 'religious' systems. As with the negotiations of the Indigenous Medical Bill, the designation of areas as Hindu or Muslim, and the consequent assignment of Ayurvedic medicine or Unani Tibb along with the appropriate practitioners to the region, worked to cement a mentality that saw the traditions as separate but equal. While this approach was out of line with the official secularist agenda of the Congress party, it was emblematic of the key principles of professionalization: the identification of practice and the demarcation of belonging. The representation of the relationship between Ayurveda and Unani as one of symbiotic co-existence but careful separation, and the expansion of services based upon this ideal, served to reify a 'historical' vision of the Indigenous Medical Systems that had actually only existed from the 1920s on and had in fact been created by the Board of Indian Medicine. However, the alleged 'historicity' of this bifurcation found legitimacy as part of the increasing communalization of politics in this period. The secularist agenda in this context became about adequately identifying community-based divisions, casting medical care as an inalienable tradition of community, and providing for the continuation of the tradition. The non-urban dispensaries had been, in some ways, the last vestiges of hybridity and cultural mixing that had characterized the consumption of Ayurvedic medicine in this region.[40] From 1938 on, as we shall see, community displaced other economic and social factors as the prime concern in the planning of medical care.

Until 1938, investment in the public health of those citizens outside of the urban context received public assistance only in times of crisis, for example during the famine or health epidemics. Everyday healthcare practices in those regions were completely separate from the state, even when situated within prominent institutions. The investment in rural dispensaries took great steps towards reversing this arrangement, in part by absorbing already existent institutions into the state framework by granting them status as official state-aided establishment, and partly by identifying areas were institutions needed to be built. This latter move proved to be politically significant: for in so doing, they pushed the boundary that divided the Medical Board from the Rural Development Department, absorbing responsibility for the everyday practices of populations considered to be on the margins of political life by incorporating them into the political mainstream. To seek medical care at a state-aided *Aushadalya* or *Dawakhana* was to participate in the formation of nationalist political identity, to ostensibly be part of the nation of political subjects governed by elected officials.

## The limits of governance: Educational institutions and the indigenous response

One of the most significant consequences of the government's decision to expand its recognition of indigenous medical systems and to incorporate them into official state medicine was the establishment of educational institutions of Ayurvedic and Unani learning. Unani institutions have a very different history from Ayurvedic ones that began during this period. As Seema Alavi and Guy Attewell have shown, Unani profession had begun to assert Islamic identity within North India.[41] Ayurvedic practitioners professionalized much later on. As we saw in the first chapter, the institutional affiliation that vaids possessed derived mostly from their abilities as pandits versed in Sanskrit, and familiar with the study of ancient texts. Despite the presence of Unani practitioners, and despite its emerging importance in the public sphere, Ayurvedic study occupied no comparable position in official state discourse until the 1920s, when the Board of Indian medicine developed criteria by which Hakims and vaids could be tested. Unani practitioners, ranging from Ajmal Khan in Delhi to local family firms in Lucknow, had begun an informal process of standardization as early as the 1880s. At the same time, hakims had consciously formed professional networks that mediated medical authority within the major cities of the United Provinces, linked in part by the intellectual elite based in Aligarh, but equally important to the functioning of the region's metropolitan centres.[42]

The professionalization of Ayurveda occurred much later. As late as the 1920s, the primary method of teaching Ayurvedic skills was through the *guru–shishya* model. Though the teaching of several students at once was not uncommon, constituting a classroom setting as opposed to the apprenticeship model invoked, the quality of teaching relied solely on the competence of the individual teacher.[43] The authority of certain vaids derived mostly from their acceptance in the community and was determined predominantly by their ability to provide care. Certain *Aushadalyas* and clinics were famous throughout the region, and, in one case, throughout the subcontinent.[44] Though this provided cultural and social currency for the vaids which resulted in economic profit, this did not conform to the new government regulations outlining the official role of indigenous medical practitioners in the UP. The Board of Indian Medicine insisted on a process of institutionalization that would theoretically ensure that a similar standard of training had been received by vaids throughout the subcontinent.

Though countless institutions of Ayurvedic and Unani learning existed in the UP, only the BHU faculty of Ayurveda and the Unani College at Aligarh were initially recognized by the government as being of the first rank, and the government considered all others in relation to them. While funding was given to family-based firms like the Takmil-ul-Tibb in Lucknow, it was the University faculties that were employed as models of appropriate education.[45] The BHU's Ayurvedic College was founded in the early 1920s, some seven or eight years after the foundation of the University.[46] The founding of an Ayurvedic college has been one of the aims of the University's founders, but it was only after the BIM's grant of Rs 50,000 in 1923, and promise of an annual recurring grant of Rs 2500, that the Ayurvedic department could become its own College, replete with attached hospital. The course taught there was the most rigorous Ayurvedic course in the Province, encompassing six years of study, after which successful students could register as Class A practitioners, and claim the indigenous moniker of *Ayurvedacharya*. The government considered BHU graduates to be the best trained in the province, and often selected them for related government jobs, as we shall see. BHU graduates were certainly the most professionalized of the Ayurvedic students, situated as they were at one of the great modern institutions of the UP, which was itself at the forefront of the political modes of reforming education.[47]

Not all of the proposed institutions, however, fell in with the trappings of modernization with which the BHU actively engaged. This was particularly true of the Rishkul Ayurvedic College in Haridwar. It opened in the early twentieth century as a centre for Vedic learning, and in 1926 petitioned the government for the ability to grant certificates in Ayurvedacharya, which would allow its students to claim status as Class B practitioners.[48] The category of Acharya, a tradition rooted in Sanskritic tradition and not in the lived experience of practice, was the main point of contention: the Institute insisted upon contextualizing the teaching of Ayurveda within the more traditional aspects of practice, stressing the authority of the Sanskritic past in determining its boundaries, and removing the connotations of medical 'modernity' within which the BIM had been trying to frame the tradition. The funds for the college were privately organized, collected through tuition and donations, and the Medical Council could therefore not restrict the curriculum of the college. However, it could restrict the official legitimacy of the college and its practitioners. Rishkul Ayurvedic College, therefore, makes evident both the scope of reform that the Board of Indian Medicine had and the limits to its reach.

The curriculum of the College was emblematic of other state-aided institutions that produced Class B practitioners. The curriculum was divided into allopathic and Ayurvedic learning, completed over the course of five years.[49] The general science model included the study of physics, chemistry and biology, taught in the first and second years of study, and the rest of the degree was dominated by aspects of the Ayurvedic tradition.[50] The teaching staff of the college hovered around 10 or 11 members, most of whom were vaids, along with an MBBS-accredited doctor, and a BSc to teach general sciences.[51] The College taught the course prescribed the BIM and administered its exam, which was examined by at least one external representative of the local government.[52] There was a hospital attached to the College, as well as a basic science lab in which Western medicine was taught.[53] It had very close ties to the Rishikul Brahmacharya Ashram, a Sanatan Dharm institution, where Sanskrit was taught to the vaids, despite the Ashram's complete dissociation with the education department.[54]

The initial funding for the Rishkul college had come from Lala Sukhbir Singh, a regional monarch, who then went on to represent the college in its dealings with the government.[55] After some deliberation regarding the scope of the school and its potential importance as a central institution, the government's initial response was to classify the College as a teaching college that would receive a grant-in-aid administered by the BIM. The result of this would be that the practitioner who graduated form the college's Ayurvedic programme could, following registration, be employed within the scope of government, but only at the level of a Class B practitioner; in practice, the practitioners, students and scholars employed at the school would most likely be marginal to the functioning of the Board and of the other bureaucratic apparatus of the BIM, which tended to draw on BHU-associated vaids to officiate on regulatory and legislative matters when an 'indigenous' opinion was required.[56] The negotiations over the opening of the college, therefore, concerned the size of the grant, and not the status of the institution, and it was decided that a recurring grant of Rs 10,000 as well as a one-off grant of Rs 15,000 would be allocated from the financial department straight to the college.[57]

The administration of the grant, however, depended on the Institute's formation of a formal constitution that would clarify the role of the government in the functioning of the college.[58] This is where the troubles began. The government imagined the constitution of the college, as well as its Board of Directors to be similar to the other college that had received funds that year, the Takmil-ul-Tibb in Lucknow. The Tibbi school's managing committee consisted of the district commissioner (who would serve as ex-officio president), the local collector, the chairman of the BIM, the assistant surgeon of the region, two vaids and two hakims, a member nominated by government and the principal of the school. This collection of individuals represented a balance of the different medical and political agendas that maintained the balance of power between Muslims and Hindus (despite the Tibbi schools obvious Islamic affiliation) and the different factions of the political scene. The Rishkul College objected on many fronts. They proposed in turn a committee made up of 11 members, seven of whom would be of their own choosing, four of whom could be chosen by the government, with the principal of the school acting as the President of the Governing Body. The college's manoeuvres in the negotiations that followed make evident that its main concern was with the number of vaids who would sit on the committee: the government proposal would have seen two hakims sit on the council as part of the board policy of communal representation. Indeed the proposal it put forth, which was finally accepted by the government, called for the principal of the school to sit as president of the governing body, and for the government to elect in its three allotted officials at least one vaid; the other members would be chosen by the Board of Directors of the school, and would include at least two vaids.[59] The government, by this point, had expressed frustration and annoyance with the negotiations, which had by then lasted two years and had delayed the deployment of these funds; the compromise was accepted and an overtly Hindu committee of governors was appointed.

This successful Hinduization of the Committee may appear a rather innocuous outcome, and the appointment of many vaids to the Board of Governors might be construed as a positive development for an Ayurvedic college of the day. By the late 1940s, Rishkul College was a prime training space for the Rashtriya Swayamsevak Sangh: incidents of riots staged and recruitment initiated by the group had been reported since the late 1930s. A report commissioned in 1949 revealed that the college, despite the regulation of government officials who sat upon its board, had always been under the control of the Rishkul Brahmacharya Ashram, run by Sanatana Dharm Hindus connected to the Arya Samaj.[60] The Ayurvedic college acted as a satellite for the Ashram, over which the government had no control. The college certainly succeeded in peopling its board with informed Indian members, and had more Indian members than any other educational institution of its status in the UP at this time. However, the attempt to protect the college from Unani influences, especially in an era where Unani colleges often included both Hindus and vaids on their committees, is cause for consideration, and the later history of the college sheds light upon the significance of these early developments.[61]

The practitioners most vocally critical of their role within the revamped professional world of indigenous medicine were those at Rishkul Ayurvedic College. In a 1940 speech addressing the first Indian

Commissioner of Saharanpur, who was at the college to lay the foundation stone for its new hospital, the principal of the university criticized
the government's tendency to predominantly employ in its rural development schemes graduates of the BHU.[62] The principal began the speech
by praising Mother Ganges, and lauding the commissioner on his return
to the temples of Brijbhumi and Haridwar following his travels abroad,
and his continued devotion to Lord Krishna. However, this had not
inspired him to look kindly upon the many achievements of the graduates of the institution, who occupied many public roles, including
'Pilgrim comfort inspectors' at Melas and Ardi-Kumbh festival, in development charitable dispensaries and in their roles as members of Vaidya
Mandalas. 'The Rishkul Ayurvedic College Hardwar is the only institution which helps its students in securing diplomas by teaching them
the course prescribed by the Board and sending them for the Board's
Examination, but whenever there arises any chance for the so-called promotion of the Vaidyas, it is not known why your Government always
gives preference to the graduates of Banaras University who neither
study the course prescribed nor obtain diplomas.'[63] The principal went
on to list the subjects for study, which included the basic medical topics, plus medical jurisprudence and female and childhood illness; he also
stressed the importance of allopathy to the curriculum, though perhaps
he unscrupulously noted that between 60% and 90% of students failed
their allopathic medical exams, which potentially harmed his chances
of having his students granted a certificate of allopathic medicine.[64]

The students, apparently, were not altogether more successful in the
BIM-sanctioned general exam, and could therefore not be registered
with the BIM, and were thus being left out of some of the more lucrative career prospects for indigenous medical practitioners. The principal
was ultimately trying to twist the arm of the commissioner by arguing
that the new rural development schemes were so reliant on Ayurvedic
knowledge that the BIM-determined standards no longer applied. The
BIM had reiterated in a 1940 decree that vaids and hakims who were not
registered could not issue certificates, do medico-legal work or perform
cholera or plagues inoculations; rural practice, the principal argued,
encompassed none of these duties:

> From a study of [these claims], it becomes clear that a simple method
> has been through out for subjecting the Ayurvedic system to ridicule
> by crippling Vaidyas. Besides, when the knowledge of allopathy can-
> not be utilized for benefiting the Vaidyas and the public, no special
> importance attaches to its inclusion in the syllabus.[65]

Moreover, Rishkul graduates who were able to pass were being passed over for posts; the principal claimed that this led to the further disillusionment with the system, which resulted in more students failing.

The chairman of the BIM, in response to the speech, which had been presented to the BIM scholars committee in written form, disagreed with the complaints lodged by the Rishkul Ayurvedic College. In terms of appointment to the Rural Development Department, the BIM admitted that BHU students were favoured as they received the highest level of training and 'A' class status, but also that a large number of 146 appointments in the RDD had gone to Rishkul students.[66] While it conceded that inspectors had been chosen from the ranks of Aligarh and Banaras graduates, it also claimed that graduates of the new five-year Ayurvedacharya programme would also be eligible to apply in the next round. On the grounds of allopathic certificates, the government refused to engage – an allopathic certificate was lodged by the college as a replacement for an allopathic medical degree, and the BIM would not entertain this measure.[67] The other complaints of the BIM scholars were not taken into consideration, and the report was summarily dismissed by the Medical Council and the Department of Public Health.

The government's take on graduates of Rishkul was specious at best. An offer by the vaids at the institution to send a delegation to the Army Services was deemed nicely patriotic, but completely useless.[68] In a more acerbic turn, Ivo Elliott of the DPH wrote to the claimed that the B.I.M. Scholars Association was itself a fallacy; claiming that the students '[sent] these request under this highsounding name, ... [but] will be entirely misfit in the medical establishment of the field service, and it will be a waste of time of the govt of India Medical Directorate to send a representative down to make any investigation'.[69] Indeed, the students and vaids at Rishkul posed a problem for the government: while adhering to some of the principles of the newly professionalized systems, they remained fundamentally rural practitioners 'limited' by patterns of practice that lay beyond the scope of legislative reform. The students at BHU and the Takmil-ul-Tibb could enjoy a regional cosmopolitanism buttressed by the experience of drastic urban change that made the 'modernization' of these systems resonate as part and parcel of an Indian modernity. Though regional centres like Haridwar were certainly affected by the changing political structure of the late 1930s and 1940s, the changes did resonate as deeply on all fronts of public life.

Despite the insistence on regulation, registration and accountability, the Board of Indian Medicine had little to do with educational institutions in the United Provinces. Government responsibility went as far as providing small grants for upkeep, renovations and one-off instances of hiring or re-stocking. While they insisted on the process of educational institutionalization as a pre-requisite for the granting of certificates, there was no official protocol for evaluating individual institutions. As was the case with pharmacies, already developed institutions proved a more attractive solution to the problem of regulated education than did establishing new programmes or institutions.

This makes evident the limited scope of government with regard to the actual functioning of educational institutions, especially in smaller urban centres like Haridwar. The Rishkul example highlights the extent to which both the rhetoric and authority of government were valued by local practitioners and institutions. The vaids at Rishkul took advantage of government initiatives to create an institution at odds with the ideology behind the reform of the Indigenous Medical Systems. Firstly, the vaids trained at Rishkul failed to meet the standards set by the government for the training or practice, and continually failed. However, more significantly, the college was itself implicated in the cultural and political workings of the Rishkul Brahmacharya Ashram, where its students learned the cultural and political workings of anti-Muslim organizing along with their general studies. The secularist agenda that fuelled the policy of the BIM had no place at Rishkul, despite the fact that it was the recipient of a large annual grant and a major supplier of pharmaceutical products to other regional dispensaries. Instead, the Institution co-opted the legitimacy of the College's affiliation, and used it to further its own partisan agenda.

Despite its apparent dedication to reforming the indigenous medical systems, the Board of Indian Medicine, even under the Congress government, lacked any real power to effect change. While methods for procuring licenses and certifications were put into place, the monitoring committee intended to evaluate the efficacy of practice failed to properly materialize. Similarly, the intention to distribute pharmaceuticals produced by the most efficient and medically sophisticated firms was abandoned in favour of patronizing systems that were already in place. The dwindling power of the BIM, and government indifference to its inefficacy, makes evident the failure of the process of institutionalization to transform the practice of indigenous medical services. The changes made only affected the services provided by vaids and hakims to government servants, and had little bearing upon the public. Similarly, the educational institutions that emerged in the 1930s and 1940s were of little concern to government; they were given small grants to cover their expenses, but could operate as they chose without any fear of official interference.

The concerns introduced in the 1930s regarding the general importance and authority of the BIM in the culture of medical practice were not resolved during the colonial period, and would emerge at the centre of early postcolonial debates about standards of medical practice and care. Instead, the 1940s saw the consistent evolution of what it meant to be a registered practitioner, a notion that evolved greatly during this decade. The advent of a widespread, engaged discussion of the meaning of Ayurveda amongst vaids themselves enabled the transformation of disparate medicalized practices into a professional system of medicine. More important than colonial government as embodied in the provincial boards was the rigidification of taxonomies of knowledge and experience that were developed in the public sphere.

The major barriers to meaningful, all-encompassing reform of the systems of Indigenous Medicine lay in the simple inapplicability of the reforms introduced by government to the lives of most Ayurvedic practitioners. The BIM never had more than 200 vaids and hakims on its register of certified practitioners.[70] Where the Rishkul vaids made claims and filed reports, other UP vaids made their cultural and social authority clear to the government by simply ignoring its protocol. The Rishkul vaids remained the most vocal opponents of the BIM's policies, as well as the ones most engaged with the letter of the law and its evolving constitution. The continued efforts of the BIM to finalize the registration procedure, however, makes evident the fundamental paucity of its reach: the members of the Board and the Select Committees set up to reform its policies constantly commented on the loopholes that vaids and hakims found to get around its rules, as well as the general irreverence for the letter – if not the spirit – of the law amongst Indigenous Medical Practitioners. Despite the predominance of the public opinion mediated in the world of print capital and the consumption of informal medical services, vaids and hakims incorporated aspects of the new state institutions into their professional lives.

\* \* \* \*

Regardless of the perceived 'failure' of these institutions to serve the needs of the provincial government, they became the vehicle for other sorts of advancement. For instance, these experiences of institution-building reveal the extent to which official and popular ideals of the nation and its health care could meet in discussions of indigenous medicine. Though explored from the perspective of official communication, the importance of public opinion is evident throughout these discussions: the deployment of the idea of a consuming public is at the heart of the discourse employed by Congress and by Rishkul Ayurvedic College, both of whom sought to use the 'authentic' Indian body and its appropriate modes of health care as the basis for its political pitch. The deployment of Ayurveda as political ideology marks its final move away from being a mere composite of ideas, and crowns its entrance into the world of the consolidated, easily deployable political trope.

Ultimately, it was this straddling of colonial logic and proto-independent imaginings that characterized the unfolding of events both vis-à-vis the survey and planning of dispensaries, and the continued evolution of the Rishkul Ayurvedic College. These institutions were colonial in name, and also in the way they laboured under a logic of government, and conformed to more rigidly entrenched policy standards, firmly linked to the colonial public health mechanism. However, their founders attempted to co-opt them in order to both fuel and sustain the emerging community of nationalist Indians in their quest for authentically indigenous resolutions to social problems. Congress envisioned the deployment of a secularist ideology through a reworking of traditional institutions into modern ones, and employed the rhetoric of 'development' in order to legitimate their manoeuvrings. Similarly in logic if not ideology, the vaids at the Rishkul Ayurvedic College attempted to use symbols of legislative authority to push their own agendas.

# 6
# Reframing Indigeneity: Ayurveda, Independence and the Health of the Future

In this chapter, we will examine the evolving politicization of Ayurveda as the era of colonial rule came to an end and Indian bureaucrats and citizens shifted from imagining postcoloniality to defining independence.[1] The Second World War wrought the mass disavowal of the possibilities for empire in South Asia, where cooperative and capitulative politics had shifted against a future in Empire, and in which nationalist politics had ceased to make possible the functional carrying out of large-scale imperial policy. The most prominent political movement throughout the war had been, of course, the Gandhian Quit India movement, which began in August 1942 as a response to Britain's unilateral decision to bring India into the war, and became the nationalist articulation of anti-colonial sentiment and action. Over the course of its three years, over one hundred thousand people were imprisoned, at times shutting down realms of governance crucial to the manoeuvrings of Empire. Most importantly, the positions that Gandhi and the Congress were able to take against Empire led to swifter and more absolute forms of decolonization.

Other dissenting movements expressed their opposition to both Empire and to the formal nationalism of the INC in differently radical ways. Communists and radical socialists committed to the internationalist ideals of materialist revolution rejected the limited scope of the nationalist agenda for transforming the systems of capital introduced by Empire, while simultaneously condemning Empire for the exploitation of Indian resources and labour. The Indian National Army, headed by Subhas Chandra Bose, led a differently internationalist charge against Empire. Founded in Germany in the late 1930s, the movement was meant to act as a countermeasure to the non-violent, but also non-transformative politics of the Congress government. The INA built its membership in the wartime Pacific, with group leaders backing the Japanese against Empire in return for the release of Indian prisoners of war into their care. The eventual march in Delhi symbolized a dissenting approach to both the Empire and the burgeoning nationalist state, with the INA proposing instead a different vision of Indian independence.[2]

The most significant politic of all, however, was the political manifestation of communal tension into party politics in the 1940s. The partition of India and Pakistan had been an articulated political possibility from the Simla meeting of 1940, despite Jinnah's personal objections to the idea or his personal political strategies.[3] By the end of the war, partition was considered to be an inevitable component of the withdrawal of Empire, finalizing both a century of divide-and-rule imperial tactics and decades of formal Indian lobbying for change and nationalist articulation. The partition plans were announced only weeks before the event, following months-long speculation and agonizing over where the line would lie between India and Pakistan. The announcement of the dividing line between India and Pakistan in the summer of 1947 was only anticipated in parts, causing one of the largest displacements of people in history, with the exodus of Indian Muslims and Pakistani Hindus across the borders to each of the new states.

The picture of politics drawn for the 1940s demarcates the larger elements of political will and negotiation and leaves less room for the workings of government that plodded along despite – or, at times, because of – the major political transformations of the day. What is often left unexplored is the focus on planning undertaken by both central and provincial governments for the impending moment of independence. In essence, this moment of transition fits neatly along the trajectory that began with dyarchy and was brought to full fruition in 1947, when India became an independent republic. The success of anti-colonial nationalism, matched with the ardent political organization of the Congress Party, created the context in the 1940s for the imagining of things to come, and as the moment of transition moved closer to fruition, the interest in planning the imagined postcolonial future coloured the less-essential political movements of the moment. The mid-1940s, in fact, were a time of great political development in the form of planning, moved in part by the new impetus around 'developing' India and buttressed by the strengthening apparatus of provincial governments.

In this chapter, we will explore an era of experimental planning on a local, national and social scale, observing the ways in which an intimate familiarity with Ayurveda itself became a designation of an attractively authentic and thoroughly modern understanding of the postcolonial Indian condition. Working through several genres of contrasted knowledge production about Ayurveda in the 1950s, this chapter reveals a plurality of stakeholders, ranging from federal politicians to state bureaucrats to educational instructors to lay authors, who scrambled to lay claim to an authoritative knowledge of Ayurveda, and who together formulated a politic of the possible for Ayurvedic medicine. We will begin at the national level, with an exploration of the Report of the Health Survey and Development Committee (the Bhore Report) and the Report of the Committee on Indigenous Systems of Medicine (the Chopra Report), the latter filling the gaps that the former identified. We will continue on to the pragmatics of change on the ground in the newly independent Uttar Pradesh, initiated through a revamp of the public health and health education schemes geared to the new preventative public health agenda. Finally, we will turn once again to the public sphere, where the topic of food had come to dominate the discourse of health once occupied by the reproductive body. What emerges from these myriad attempts to define Ayurveda is the fundamental nature of its transition in this period: similar to all stakeholders was the understanding that the system to which they were laying claim had succeeded in becoming sufficiently couched in the language of techno-science so as to sustain its position as a cornerstone of state-building, while simultaneously relying upon the authentically indigenous systems and processes to provide immediate redressal to the pressing problems of independent modernity.

## Between Bhore and Chopra: Techno-science plotted and deferred

Having weathered the storms of several elections by the mid-1940s, the Congress Party had begun to flex its political muscles by envisioning possible postcolonial futures crafted with commitment to the party ideals of secular, liberal and global modernity.[4] This schema was seen across a range of departments and themes, but was possibly most thoroughly articulated when mapped across the bodies of the citizens of India in the 1946 Bhore Commission Report, undertaken by the Health Survey and Development Committee. The Bhore Report, as it came to be known, revealed the results of a three-year epidemiological survey of the health of the Indian populace, an unprecedented hallmark in Indian population studies and public health. The report is best known for its attempt to map a national health plan for an imagined post-war – and independent – future.

The Bhore Report has distinguished itself historically as the first major attempt to approach the question of Indian health on a national scale, and from a solely techno-scientific standpoint. The commission was composed of a team of leading Indian and British civil servants and medical consultants well-versed in the intricacies of international discourses of population threat (and subsequent control), the spread of diseases and the global spectre of local problems.[5] The evaluation of India's role as a global actor within spheres of health was measured in relation to statistics gleaned from other localized studies as well as statistical data collected by the League of Nations.[6] India was placed alongside the United Kingdom, Canada, the United States, Australia, New Zealand and, due to the communist Zigerist's contribution, Soviet Russia.

The report was primarily intended to be a road map for thinking through the matter of health as a component of state planning, and was perhaps most successful at setting the tone for other endeavours by thinking bold and thinking big. Its value to historians lies in the framing of this moment of transition from the point of view of political transformation – it was not a response to a crisis, or an attempt to radically conform or break away from imperial politics. Instead, it aimed to define the politics of the future in cautious and careful ways, asserting an independent future but with great heed paid to the political constructs of empire and global neighbours. The language used to distinguish the past from the future hinged around the notion of the war: references were continually made to the pre-war period, a category that still governed the warring present moment of 1943. The future was signified as post-war, tying the independent future of the subcontinent to the end of the war. At the same time, the notion of 'wartime' was invoked to explain and excuse the limitations of the study, which were meant to be preliminary guidelines.

The post-war future, in which India's then-precarious independence would be won, was also invoked in discussions of the scope of the study, especially as articulated by the question of cost. The expenses were a concern both pragmatically and hypothetically, as the exceptional conditions of wartime precluded the kind of spending the committee would have liked to invest in this kind of survey. A healthy future for India was hinged upon the idea of an increase in health spending, and the postcolonial was characterized by a government commitment to health. In the words of the report, the commission was urged to 'to plan boldly': it was urged to avoid extravagant programmes that would never be imaginably feasible, but also not to engage in 'halting and inadequate schemes which could have no effect on general health standards and which would bring little return for the expenditure involved'.[7]

The report began by defining the notion of health for the purposes of the study, and subsequently, for the purposes of governance. Health, accordingly, 'implie[d] more than an absence of sickness in the individual and indicate[d] a state of harmonious functioning of the body and mind in relation to his physical and social environment, so as to enable him to enjoy life to the fullest possible extent and to reach his maximum level of productive capacity'.[8] The population was accordingly split along the axes of the resulting taxonomy: those whose 'levels of health is so low that they are victims of disease; others who do not manifest symptoms of disease, but are devitalised and face restrictions to their physical and mental achievement; and, finally, those 'blessed with an abundance of life and vigour'.[9] Due to the limits of the statistical data available, the committee remarked upon the inability to enumerate the good health – represented in the latter category – of the nation, and instead staged its reading of ill health through statistics on mortality.

The attempt to coalesce the different factions of the diseased nation into a coherent whole reveals the fetishization of the concept of population at work in the report. Where population discourse had developed significantly within the contexts of the local and the imagining of the international, the GOI had defrayed attempts to address the question of India's teeming masses and the threat they posed to the workings of governance. The Bhore Report's significance was its attempt to conceptualize the management of the national population as a cohesive whole. The enactment of the plan saw the re-conceptualization of space for the purposes of population management. The colonial division of districts into *tehsils* (subdivisions) was based on land and was replaced in the Bhore Report with a plan to redistribute resources on the grounds of population density rather than established regional boundaries.[10] The reimagining of the landscape based on the demands of population effected the major resolution of the report, which was to thoroughly modernize the practices by uniformly ensuring that all deployed health care come into some contact with the apparatuses of a biopolitical modernity.

A three-tiered assessment of the public health lay the grounds for recommended interventions. India's high mortality rates seemed to originate at a foundational level from the twin dangers of unsanitary living conditions matched with malnutrition, the latter characterized by both a lack of food and the inadequate caloric value of the food available. Following these problems, the inadequacy of medical and preventative health organizations already extant resulted in an inability to stem the tide of illness and disease stemming from unsatisfactory living conditions. Finally, an overall dearth of general and specific health education was identified as the extenuating factor that contributed to the condition of ill health that plagued the nation. This last factor brought to light some of the mitigating factors that caused overall social disease, which the committee argued resulted in tangible health concerns. The practice of purdah, associated in this report solely with the Muslim community, and the practice of early marriage were both indentified as main contributors to infant and maternal mortality, alongside generally widespread illiteracy and apathy towards conditions of sanitation – an argument that mirrored precisely the interwar framing of gender and sexual relations as impediments to progress and framed reform measures.

The overall strategy of the report built keenly on these twinned approaches, highlighting preventative medicine as the key to sustained public health, while building in strategies of surveillance to ensure its practice. The gist of the plan for rural India was to train key practitioners and other local health overseers in the basic tenets of medical modernity, who would then act as intermediaries between the state and the masses. These 'medical officers' would be primarily trained in preventative medicine, with the aim of ensuring that the threat that everyday practices brought about disease would be addressed locally and then reported on to the appropriate state representative. At the same time, the prominence given to women's issues brought the private sphere of domestic life into the realm of the social, mirroring well-established practices associated with the gendering of health in the private sphere. By singling out purdah, marriage and midwifery as major areas of change, the state was able to extend the reach of its influence into the most intimate spheres of Indian life.[11] While the role of the state in the intimate space of the family was neither a new nor an uncommon practice, to frame gendered practices as a foundational cornerstone of social and political life in such an obvious and unbracketed fashion was itself an interesting innovation.

The report also broke ground significantly with the incorporation of a different kind of spectrum of private life into its plans. Where gender and sexuality had cut through reforms from the early days of colonization, sometimes incorporated and at other times bracketed off into highly specified categories of concern, the pragmatics of local health practice had been consistently relegated to the parameters of the household or, at most, the village. While the reform of the indigenous medical systems in the 1920s–1930s had established a cohesive category in which these practices could be conceptualized, they had also been strictly relegated to the confines of provincial politics and redirected away from the purview of national governance as such. The focus in the Bhore Report on local and everyday practice as a cornerstone of national health implicated the practice of extra-allopathic medical practices as a necessary part of national health planning. The commission, however, had very little to say about the Indigenous Systems of Medicine – save their popularity and economic efficiency, they were deemed to be unquantifiably measurable in terms of the impact of their value and were cordoned off for further discussion.

This paradox of bold plans on tight budgets is the fantastical legacy of the report, which lived on in the historical memory of planning as a model for managing and improving the overall health of unwieldy populations. It also managed to remain the authoritative study and set of proposals on the theme of health in India until the end of the twentieth century, with scholars relying on its aims and its approach to scale as late as the mid-1990s.[12] Ultimately, however, its far-reaching suggestions for a complete transformation of the infrastructure and pathos of Indian health care proved too ambitious for the pragmatics of health governance, and its suggestions were widely left by the wayside as new plans were drawn for dealing less dramatically with the problems posed by the rural poor. Its legacy, though, is in its adherence to a techno-scientific model for dealing with unruly subjects, namely 80% of the population living beyond the scope of the state's biopolitical reach. Ultimately, the harnessing of this vast segment of the population into the planning ideals of the Indian future worked to resituate the rural poor within the framework of the nation-state.

### For the benefit of humanity, a unified system for the country: The Chopra report, 1948

Previous chapters have delineated the extent to which Ayurveda was developed at the provincial level, and more meaningfully articulated in regional, urban and suburban contexts as a restructuring of health governance in emergent political spaces. At the centre, however, the indigenous medical systems had failed to gain any sort of similar ground: while the shrinking IMS maintained an interest and a guiding arm during moments of epidemic, the subject of primary, preventative medicine had been side-lined into questions of hygiene, which were best articulated in other spheres. Historians have delineated the Raj's ideological shift in thinking about medicine and disease as moving from a nineteenth-century fear of contamination to a twentieth-century concern with contagion, which was made manifest in the invented category of social hygiene.[13] While this would seem to be the work of the Indian Medical Services, the imperial shift to a broader ideology seemingly focused on the 'art of governance', coupled with the introduction of diarchic devolution, resulted in the articulation of the social hygiene concerns across a wide sphere of political portfolios and colonial (or para-colonial) institutions.[14] Social hygiene was more integral to the realm of urban planning than it was to the functions of the diminished Indian Medical Services, for instance, and the broader questions of population control and management were more profoundly taken up in lay discussions of sexuality, as we saw in Chapter 3. At the same time, the burgeoning field of sexology posited Indian experts, who laid claim to both Western scientific and Indian indigenous knowledge of the body and its functions, as new authority figures on the body and its social dimensions in the populous. Many of these sexologists wrote for international audiences and participated in cosmopolitan conversations about the science of sex and sexuality in the international sphere.[15]

The context of preventative medicine, therefore, had moved well beyond the realm of the IMS. However, as the Congress took the lead on questions of governance, the topic of prevention was reintroduced back into the lexicon of grand-scale state medical planning. As it had in the provinces, it found its early independent articulation in relation to the pre-extant contexts of Indian health – namely, the Indigenous Systems of Medicine. Throughout the interwar period, various attempts had been made to assess the state of the indigenous systems for the purposes of a central government: most notably, with the advent of the 1923 Usman report, then mirrored by the INC's Working Committee Document of 1938, and finally followed up with the far-reaching but impractical Bhore Committee Report. None of these reports resulted in anything more than the establishment of a discursive category of care-speak, fitting neatly into the broader discourse of development employed by the Congress as it addressed the needs of the teeming masses now under its jurisdiction. As such, the provinces remained autonomously invested in the elaboration of the indigenous medical systems within the logic of local statecraft, and succeeded in far surpassing the central government's weak attempts to determine the role of the indigenous medical systems within the structure of governance.

The pattern was slightly reversed in 1948 with the publication of the Report of the Committee on Indigenous Systems of Medicine, chaired by R.N. Chopra (and thus securing its unofficial moniker, the Chopra Report). The idea behind the committee, as was the case with the others, was to investigate the state, relevance and applicability of the indigenous medical systems to the health challenges facing the subcontinent. The committee was also tasked with bringing the Indigenous Systems of Medicine into line with the values around medicine, disease and, ultimately, health, which the government was espousing. Where other reports had taken great pains to differentiate between traditions, linking consumption to communal identity (and, rather broadly, linking medical systems to different types of indigenous bodies), this committee was tasked with the goal of rationalizing Ayurveda and Unani into one singular system, and thus erasing the distinction between them. While the first few pages lay out the agenda of the report, Chopra spells out this task at the end of his introduction in no uncertain terms:

> We wish to emphasise here that the Committee do not believe in the multiplicity of systems of medicine. Science is universal and medical science is no exception. The so-called 'Systems' merely represent different aspects of and approaches to medical science as practiced during different ages and in different parts of the world. Anything of value emerging from these should be utilized for the benefit of humanity as a whole without any reservation and integrated in the form of a unified system for the country.

The first mark of distinction between this and other attempts to plot a nationalist medicine, therefore, was the deployment of a Nehruvian secularist approach to the realm of the traditional, legitimated as a fragment of modernity through the idiom of rational science that it was meant to be able to reveal. At the onset of the investigation, the committee's aim – weighed down heavily by the concerns of the nation in wait of good science – was to determine whether the indigenous medicine was, in essence, 'modern' enough to transcend the communal, caste, regional, linguistic and other divisions that rendered 'tradition' a part of an antiquated past – and thus not suitable for the trappings of modern governance.

As often belied the project of secularizing the socially entrenched workings of Indian politics seemed to fail by its own yard measure from the onset of its commissionary *raison d'être*. On the first page of the report, the authors spelled out a history of Ayurveda's indigeneity, identifying it as Vedic practice (balanced out over four stages of development, spanning 3,000 years), and in contrast casting Unani Tibb, the other indigenous medical system under observation, as both foreign and a recent find, having been introduced in India only 1,200 years earlier. Despite this striking difference in historical categorization, the goal of the committee remained the easy reconciling of one tradition with the other.

This report differed from others, however, in its attempts to spell out the terms of its historical development, as established through the textual history of each tradition, its inevitable decline and its eventual revamp. Ayurveda's beginnings were traced back to Vedic times and made relevant as the medical teachings available to the Aryans, during which time treatises on medicine, anatomy, gynaecology, obstetrics, caesarean section and crushing of the foetus, surgery, lithotomy and rhinoplasty were composed.[16] Central to Ayurveda was the *Tridosha* system, which the committee claimed could reconcile two antagonistic theories, namely the cellular and the humoral, the latter of which had been much maligned within the framework of modern medicine. The committee also described Ayurveda's spread to Egypt, Greece, Rome and Arabia, and made lofty claims about its influence:

> Through its influence on Greek medicine and through the influence of the Greek and Arabian medicine on medicine in Europe, the Ayurvedic system can well claim to be the chief, though remote source, from which the mighty river of Western medicine has had its beginnings.[17]

Despite these lofty ancient and medieval pasts, Ayurveda was still not immune to periods of decline. Interestingly, the committee allots this in part to the growing discomfort amongst the Hindu community with the dissection of bodies, and the eventual emergence of cataract procedures, bone setting and the surgical treatment of the wounded in battle as the major highlights of Ayurvedic development.

Unani Tibb had a different historical trajectory linked to the political victories of Islam. Named to commemorate the influence of the Greeks on Arabian traditions, and linking the emergence of medical learning to the University of Baghdad in the mid-eighth century, Unani is said to have flourished under the pen of the tenth-century writer Avicenna (Sheikh Bu Alisenna) who wrote his Qanoon, which was then translated into a variety of learned languages.[18] Unani's presence was felt in eleventh- and twelfth-century Spain, where scholars translated great works of the tradition into Latin (though the Qanoon would not be translated until 1593). The decline of the tradition is linked to the end of

Islamic dominance in Europe and the committee's stagnation of Arabic learning.[19]

The nail in the coffin of development in both traditions was the onset of colonial rule in India. The committee report begins with a genealogy of government attention to the indigenous medical systems from the early nineteenth century, citing the development of classes in Ayurvedic medicine at the Government Sanskrit College, which were, of course, put to an end in 1833.[20] The commission identified some movements beginning with the formulation of the All-India Ayurveda Mahasammelan (ABAM), as well as the subsequent attempts by government to solicit 'state of the field' reports on the organization and function of Ayurveda in rural society. A neat line was drawn between the opening of provincial colleges, the constitution of provincial committees, the Congress Working Committee document of 1938, and that came to rest with the Bhore Report. In so doing, the Chopra Commission revealed the pinnacle moment of development from which the Chopra Commission received its charge: beginning with a quote from the Bhore Report that recognized but ultimately dismissed the Indigenous Systems of Medicine, the commission was clear that its mandate was to assess, explain and improve upon the systems of medicine that served '80% of the population'.[21]

The representation of a dearth of development between the early nineteenth century and the mid-1940s set up the political scope of the Chopra Report quite neatly. In practical terms, it reified the undertaking of the Bhore Report at the precise moment when it faced its harshest criticism both nationally and internationally for being too lofty and expensive, while responding to these criticisms by offering pragmatic solutions that held on fast to Bhore values. At the same time, it allowed for a smooth entry point into a thoroughly Indian modernity, characterized by attempts at planning, development and techno-science, rather than the trappings of the colonial modernity of the Raj – the report asserted a Gandhian ethos articulated through a Nehruvian vision of India's future.

Despite the affective framing of the report within its historical context, it was conceptualized as a pragmatic response to the Bhore Report and attempted to offer an economically feasible and strategically advantageous set of solutions to the challenges presented within it. The initial problematic remained the same: that Indian health care was predominantly 'backward', that India's health results and outcomes were estimated to be the lowest amongst regions surveyed and that the 80% of those living in rural areas fared worst.[22] This last statistic provided a seemingly contentious politic that the Chopra Commission needed to overcome, and from whence it would draw its *raison d'être*: if the main problem with India's health conditions lay in the practices of its rural population, and this 80% turned to indigenous health practices to address their ills, then why resurrect and modernize these systems? Bhore had indeed come to this conclusion; Chopra, however, used this veiled accusation to paint a picture of the Indigenous Medical Systems that distinguished them from the category of everyday practice, formalizing for the national government a process that had long been under work, but which had, until now, escaped the purview of governmental discourse.

The Chopra Report thus offered an assessment of the state of the IMS through a series of enumerations and other attempts at scientific description that would line up with those of the Bhore Report, but provide a counter to the stark dismissal of rural/non-allopathic health conditions. Where Bhore had identified almost 47,000 doctors of whom only 13,000 were registered, the All-India Medical Conference identified 1,09, 600 known practitioners, of which 51,700 were registered, and estimated that the number could be as high as 2,00,000 to 5,00,000.[23] Similarly, there were 51 hospitals and 4,000 dispensaries that were registered, which together served over almost 15,000,000 people, with over 51 teaching institutions set to increase the number of people served by the system. It was clear numerically speaking that the Indigenous Medical Systems had the reach to create large-scale health changes on the ground.

The remainder of the report built off this latter premise by suggesting different forms of change and transformation that might be undertaken. The largest of these would be what the committee termed the 'synthesis of medicine', which would bring Ayurveda, Unani Tibb and Western Medicine in line in order to 'increase the usefulness of the systems to the public as part of a comprehensive plan'.[24] The goal of synthesis, however, was easy to argue for, but almost impossible to conceptualize, and suggestions were few and far between. The committee lauded the holistic assessment of disease in Ayurveda and Unani as something that Western medicine might take into account, seeing within the *dinicharya* (daily routine as outlined in the *Susruta Samhita*) and what they termed the 'soil factor' of Ayurveda a precursor to public health planning. At the same time, the committee advised that the absence of a scientific method in Ayurveda or Unani might be waylaid by the presence of a guiding philosophy at the heart of the system, with analogies made to Sir James Jeans and Wilfrid Trotter's ruminations on the centrality of cosmic philosophy and the importance of wonder and curiosity to the furthering of scientific innovation in biomedicine.[25] The committee offered several platitudes around knowledge exchange and assimilation in ancient Ayurveda, including the long-favoured and oft-repeated assertion that Ayurveda formed the basis of ancient Greek – and hence both Arabic and contemporary Western – medicine, along with several proclamations by medical officials speaking in favour of synthesis. More compelling, however, were the alternatives to synthesis offered by the committee, which advocated for borrowing between traditions rather than a radical transformation of either, an approach that accounted for the widespread use of penicillin, quinine and other biomedical techniques and technologies amongst Ayurvedic practitioners.[26]

The compromise between synthesis and borrowing was a proposal to revamp the educational structure for medical education in the subcontinent to reflect Ayurvedic and Western medicine (though, notably, with no specific incorporation of Unani medicine), and to tie future medical research projects to the principles of each system. This proposed program of integrated studies would streamline both medical traditions into courses that could be taken concurrently, taught by instructors trained in each tradition. This style of teaching would gradually evolve to be taught by a single instructor, trained in both traditions, using a textbook that incorporated both traditions. The final step would be the creation of research institutes where both traditions were incorporated into an integrated research methodology.

The report returned to more pragmatic and immediate concerns vis-à-vis the public health in its culminating sections on preventative medicine in rural contexts. The suggested plan of action was to create, as Bhore had suggested, a cadre of trained public health officials to address the needs and concerns of rural populations, through primarily preventative medicine. A six-month course in rudimentary public health issues was proposed for those with no formal medical training but with some involvement in the health care; midwives, for instance, or the assistants of vaids and hakims; those with a degree in the ISM could forego the course and take an exam instead.[27] Eventually, these trainees could oversee the *tehsil*, district and provincial hospitals and work in more strategically administrative positions.

## Reorganizing Ayurveda and Unani in Uttar Pradesh

One of the final sections of the Chopra Report dealt with the issue of national versus regional registration, arguing strongly for the creation

of a central register of authorized practitioners. This call for nationalization stands out amongst the various suggestions of the report, as it
is one of the only variations on the steady representation of the state
of the field. While the commission was organized around the idea that
the Indigenous Systems of Medicine needed transforming and modernizing into a streamlined medical practice, the actual changes proposed
for the national organizing reflected processes that were already well
entrenched in the provinces of British India. Ultimately, the Chopra
Report's plan of action mirrored to a tee the plans that had been in place
at the provincial level since the dyarchic restructuring of the mid-1920s:
the modernization of education; proposals for invigorated research; the
drawing up of plans for surveying and addressing the public health in
extra-urban territory.

Though nationalization would have been a swift and easy process – a
mere tabulation of provincial registers and reports – the ISM went on to
remain the sole purview of the states of the Republic of India until the
mid-1970s, at which point a federal agency was created to register and
manage practitioners. What this lack of action reveals is the highly conservative nature of medical planning, characterized by the expansion of
services related to – and run by – indigenous practitioners, but along
the lines proposed decades before independence. A close examination
of the discourse of planning and the varieties of plans proposed reveals
a reliance on the stability of these institutions and a reluctance to move
too drastically towards biomedicine or towards a radical reworking of
Ayurvedic medicine in practice. The only major change on the horizon
was a stark marginalization of Unani medicine and practitioners from
norms of practice or positions of power, the culmination of a historical process made manifest here through the marginalization of Islamic
culture in the newly independent Uttar Pradesh.

Interestingly, the will to conserve – rather than to improve or
transform – resulted in the provincial (and eventually state) medical
council's reorientation of its agenda for regional development around
the continued success of Ayurvedic institutions. The state commitment
to Ayurveda was expanded exponentially in terms of new institutional
and educational planning. This dedication reveals the extent to which
the architecture of infrastructure and institutions that resulted from the
transfer of responsibilities after dyarchy created the conditions for a
functional and stable political structure within the provinces of British
India, in some cases strong enough to create the conditions for smooth
transitions into postcoloniality. Historians have often argued that the
conditions for stable postcolonial development were a product of the

continued reliance on the liberal institutions of the colonial state.[28] However, the United Provinces of Agra and Awadh's transition into the Uttar Pradesh tells a different story: while British institutions may have helped the meta-structures of the state retain their solidity, it was the post-dyarchic institutions that proved stable enough to allow for a smooth transition, while at the same time being elastic enough to embrace the vision of independence.

## Defining the public health in rural UP

In 1949, a major report consisting of five separate thematic sections was issued by a body entitled the Ayurveda and Unani Tibb Reorganization Committee for the Public Health. In essence, the titling of the report marks a transition but also a confirmation of the trajectory taken by the Indigenous Medical Systems into the realms of modern health practices: rather than providing mere curative or preventative responses to the immediate medical needs of the population of Uttar Pradesh, this report confirmed Ayurveda's full entry into the biopolitical of securing the health of the nation. As we saw earlier, Ayurveda's biopolitical career took into account questions of place, space, population management and social differentiation through concentrated questions around the health of new citizens. This committee pushed the integration of Ayurveda and other systems of medicine into the biopolitical scheme of Uttar Pradesh health planning by reassessing the organization of Ayurvedic institutions, and, as we shall observe, the institution of Ayurveda itself, by bringing it centrally in line with biomedicine and its apparatuses in the state.

It is telling that the first recommendation of the report dealt with the fundamentals of institutional planning through a discussion concerning the relationship between Ayurveda and the broader apparatus of the public health.[29] The primary recommendation was for the integration of social medicine, public health and environmental hygiene into the curriculum, and that these topics should be given more time than the curative medical topics privileged at Ayurvedic and Unani colleges. The idea was to emphasize the role of these topics in the healthy organization of daily life, so that new physicians turned out by the colleges 'may be able to guide the people in their respective areas in reforming their habits on healthy lines and in improving sanitation of their locality'.[30] In the Ayurvedic curriculum emphasis would be placed upon the subjects *dinacharya, ratricharya, ritucharya, ahar* and *vihar*, all of which took into account the nature of the relationship of the body to its environs, be they domestic or public, intimate or social.

Interestingly, these courses were to stand out from others offered in the programme due to the nature of their instruction. Where other subject matter was taught by certified (or otherwise expert) practitioners or scholars well versed in the complexities of the tradition, these new courses were to be taught by district public health officers, contracted by the University to teach on an hourly basis, for which a state-wide compensation package was provided.[31] This served the public health department by ensuring an element of universal regularity along the lines of the topics being addressed; the public health officer was instructed to deliver a standard lecture, which instructors from the school would then make relevant to the students by comparing the principles addressed to similar notions within the Ayurvedic or Unani systems. Students across Uttar Pradesh would gain similar knowledge and access to information about a standardized understanding of the public health, with localized and individualized readings of the Indigenous Systems of Medicine's approach to similar matters.

In essence, the syllabus would break down into several categories pertaining to the broadest strokes of the public health:[32]

1. *Health* – definition and conception of health, promotion and preservation of health as a fundamental human right; effect of heredity and environment on health; inherited traits; healthy environment.
2. *Personal Hygiene* – cleanliness, clothing, sleep, rest, exercise, recreation, health habits; care of health in tropics; health consciousness.
3. *Ventilation, Air Conditioning and Lighting* – fresh air, humidity, vitiated air, air temperature and movement; natural and artificial ventilation, comfort zones and air conditioning; overcrowding; tropical climate and its effect on health; natural purification of air, examination of ventilation condition, airborne diseases.
4. *Dwelling and Workplaces* – site, aspect, damp-proofing and rat-proofing; residential houses, workshops, schools, hospitals, cinemas, hostels and restaurants, slaughterhouses, dairies, markets.
5. *Water Supply* – sources, impurities, purification; public water supplies; physical, chemical, bacteriological characters of good water; waterborne diseases.
6. *Food and Nutrition* – food requirements, common food products, inspection of food stuffs, adulteration, food storage, food deficiencies and nutritional disorders, balanced diets, nutritional survey, food-borne disease.
7. *Wastes* – collection, removal, and disposal of refuse, sullage and sewage; conservancy and water carriage systems; utilization of wastes, composting, sullage and sewage farms; disposal of bodies.

8. *Camp and Rural Sanitation* – fairs, camps, medical and sanitary arrangements; rural sanitation; diseases dues to insanitation.
9. *Preventative Medicine*

   a) infection, resistance and immunity. Reservoirs of infection, modes of transmission of infection and portals of discharge and entry of infection; gross infection collection and dispatch of samples to laboratories
   b) general principles of prevention and control of communicable diseases. Notification, recognition, segregation and isolation, quarantine, treatment of carriers, placarding, disinfection, disinfestation, immunization, treatment of cases and specific chemotherapy; general environmental and traffic sanitation; railways, trams, buses and other conveyance
   c) epidemiology and prevention of common communicable diseases of tropics with special reference to plague, small pox, malaria, kala azar, filariasis, diphtheria, tuberculosis, typhus, rabies, leprosy, venereal diseases, helminthic infections, enteric fever and dysentery; animal diseases transmissible to man, insect-borne diseases; insect control (flies, mosquito, louse, fleas, ticks, sand fly).

10. *Vital Statistics* – population, registration, birth, death and other morbidity and mortality rates; collection, compilation and tabulation of statistical data; common statistical error.
11. *Public Health Organizations*

    a) medical and public health organization in Centre and Uttar Pradesh, World Health Organization
       NON-OFFICIAL ORGANIZATIONS
    b) maternity and child welfare
    c) school health work
    d) industrial health organization, occupational hazards, health safety and welfare of workers
    e) health education
    f) medical and public health institutions and their administration, clinics, dispensaries, hospitals, sanatoria, health units
    g) important public health regulations
    h) health insurance and other welfare or social security measures.

These topics were far removed from those prescribed in the 1930s for Ayurvedacharya or Unani Tibb education, and were more in line with the training received by members of the Indian Medical Services and those trained for MBBS degrees in allopathic medicine. Their inclusion signalled a big break from the past, and brought Ayurvedic and Unani Tibb practitioners into the broader streamlining of the health services along biopolitical lines. The aim of the Social and Preventative Medicine syllabus was taught from an introductory perspective and in scant detail, as detailed analysis of health was not thought to be required of a 'basic medical man'. Practitioners-in-training were meant to absorb the framework of social and especially preventative medicine, and to filter through it their more detailed and precise study of the Indigenous Medical Systems. The government was careful to note that students were to be trained in social medicine because of its importance to the functioning of a social welfare state; students were trained to provide qualified personnel trained in the 'art and science of promotion, preservation and restoration of all and to remove human suffering'. The government's wording belied its fear: that students too well-versed in the particulars of public health would attempt to participate in the commercial economy of health, made up of entrepreneurs peddling the varieties of knowledge, products and technologies that we encountered in Chapter 3. Practitioners in training at colleges were to have the message reiterated to them that they were being trained as 'basic medical men' and were not to engage in commercialized medicine.[33]

By the summer term of 1952, these measures were implemented at four of the major indigenous medical institutions of the state, including the reopened Rishkul Ayurvedacharya and three other schools, with the Takmil-ul-Tibb (Unani Tibb College) in Allahabad representing traditional Islamic medicine. For the most part, the institutions were happy to adopt this aspect of the course and were able to find enough funding to pay district medical/public health officers (and the BIM agreed to foot part of the bill where the institutions couldn't). Several institutions attempted to use this platform given to them by the government to reform other aspects of their teaching and their funding structure. For instance, the instruction in surgery and postsurgical care was of concern to the Takmil-ul-Tibb College, and the principal there used this opportunity to lobby the government for further help with the training and funding of the course. The government conceded that the officer engaged to teach the public health material could advise on these matters as well.[34]

The impetus behind the reorganization of Ayurveda and Unani education along the lines of preventative medicine under a public health rubric was made evident in the recommendations made by the Reorganization Committee vis-à-vis the role of indigenous medical practitioners in non-traditional roles. The edict handed down by the committee was to 'consider the question of building up positive health through organizations for hygienic improvements and physical culture on Indian lines'. In assessing the overall schema that was used to govern health in the province, it was decided that a focus on personal hygiene and care, based on the Ayurvedic concept of *Swasthavritta*, might be of use in health relief campaigns across the United Provinces. The UP government proposed that 20 out of the 86 municipalities and 10 out of the 40 districts of Uttar Pradesh be put under the charge of 'Ayurvedists', who might imbue unto the populations of these regions the useful teachings of *Swasthavritta*.[35]

*Swasthavritta* (or, often, *swasthavritta*) refers broadly to the concept of health maintenance in Ayurveda. It is at once a specific term, but also one that characterizes the entire system, standing in as shorthand for the concept of balance between body and mind, through a daily commitment to the maintenance of health in the body. The government saw in the practice of *Swasthavritta* the potential to transform the public health through the practices of self-care. As D.A. Kulkarni, Health Minister (Ayurveda) saw it, 'if the doshic equilibrium is maintained in the body, it will successfully prevent the occurrence of organic diseases'. Furthermore, he saw within Ayurveda a set of specific rules to be followed vis-à-vis food, diet and conduct, and had high hopes that both literate and illiterate masses could easily understand and apply Ayurvedic principles in their daily life.[36] He proposed that at least one district and one municipality be put under the direct control of a qualified Ayurvedist by way of a trial run. Kulkarni suggested that newly created land entities – the district of Tehri-Garimal and the municipality of Rishikesh, neither of which had exercised power as formal entities under colonial rule – become sites of the first-run trial.

Kulkarni's excitement about the possibilities of Ayurvedic *Swasthavritta* for the purposes of large-scale preventative health schemes was not met with similar enthusiasm at the state level. While the Health Minister had been happy to go along with the inclusion of public health education in the training of vaids and hakims, he drew the line at fully integrating them into the grand-scale working of the public health apparatus of the state. In a lengthy reply to Kulkarni and the BIM, he outlined both his perception of a public health approach to medical training and his understandings of the limitations of Ayurveda. He began by distinguishing the difference between preventative medicine and Ayurveda, distinguishing between the latter's ancient roots and identifying the former as a fundamentally modern practice. Graphing time in preventative medicine through the development of the fields of bacteriology, of the science of allergens, by contemporary forays into the endocrine system, by psychiatry, along with the study of nutrition, toxins and the balance of minerals in foodstuff, Pande painted a thoroughly twentieth-century picture of the growth of the field.[37] Ayurveda, conversely, was relegated to the practice of 'therapeutic bloodletting and the polypharmacy of the alchemist' in a world where antibiotics and insecticides were used to attack disease.[38] The health officer, 'the guardian of public health' in Pande's view, was not only trained primarily in the sciences of bacteriology, public health chemistry, general sanitary measures and laws but was also experienced in the fields of maternity and infant welfare, basic paediatrics, the identification and control of disease carriage and campaigns against venereal disease and tuberculosis, along with the concerns of industrial welfare. In essence, the public health officer was at the forefront of medicine, and savvy in the politics of health; even an allopathic doctor lacked the expertise to effectively assess and address affronts to the public health.[39] A vaid or hakim was inherently unsuited to this all-important role.

Pande's opposition to Ayurvedic integration is interesting historically because of the rarity of his approach. While objections similar to those he raised would seem to be the norm amongst the biomedical community, they simply were not the norm amongst the politicized medical community, which saw in Ayurveda a cheap and relevant approach to the ills facing citizens of UP. Pande was able to capitalize on the workings of the central government in Delhi on similar issues, which, unbeknownst to him, had already been published in the form of the Chopra Report – a document which was making the rounds in circles well more powerful than the ones in which he was included, and which made recommendations in direct contradiction to his own. Historically, Pande was at odds with the medical community in failing to recognize the savvy and economically proficient use of indigenous practitioners across the frontlines of medical care nationwide.

Indigenous medicine, and particularly Ayurvedic medicine, had been moving to the forefront of frontline medical care from the late 1930s, initiated with the Congress survey and action plan in 1939, as we saw in the last chapter. The post-war, postcolonial planning agenda held to the Congress' commitment to rural development and the expansion of the public health and continued to map the population through its health needs. While Pande and others like him might have been dismissive of the incorporation of Ayurveda and Unani into formal public health apparatuses, in practice, the indigenous medicine remained the default mechanism for dealing with the health concerns of the vast majority of the population. By the 1940s, medical policy in Uttar Pradesh shifted course to formalize this practice by committing to the development of over 115 new dispensaries within the UP.

At first, this plan had been hatched to take advantage of the central government's focus on rural development; grants from the development program run by the GOI's Department of Rural Development had helped to open almost 200 dispensaries between 1947–1949.[40] However, even when it was had on good authority that the subsidies were being reduced, the Uttar Pradesh Medical Department calculated the rural dispensary plan as the cheapest available option, and hatched an ambitious plan to open 50 dispensaries between December 1951 and March 1952 (coinciding with the 1951 fiscal year).[41] This is likely due to the prominence now given to the institutionalized systems as a key to the holistic medical plan for the region – rather than cordoning off the indigenous medical systems into the schemas set up to professionalize, legalize and modernize them, the Ayurvedic and Unani dispensary plan was, in 1951, at the centre of medical planning for the region. This transition from a marginal or rural medical quick fix was confirmed when the indigenous medical dispensary began to be discussed as a model for urban development as well, moving it firmly into the centre of the cultural and political fabric of the province.

Indigenous dispensaries were favoured by government because of their cost-effective nature, which stemmed primarily from the continued take on Ayurveda as a fundamentally ephemeral, local and unregulated practice. For instance, while provisions could be made for a building and its furniture, the cost of allopathic instruments were nonexistent, as practitioners were best placed to provide their instruments themselves. Similarly, the tonics, mixtures, tinctures and other medical specimens used in Ayurvedic medicine could be sourced locally and need not come from a government-sponsored drug factory. Again, the cost would be shouldered locally by patients and practitioners alike. The government framed this in policy discussions as an extension of the local nature of Ayurvedic practice, encouraging the local cultivation and use of drugs.

Dispensaries, as we have seen earlier, also offered much more to government than a mere centre from which to distribute drugs or medical care. The mere act of their creation called for a survey of the area around them at a cost to the local region asking for help, which in turn became a cost-efficient measure for the state government to measure the demographic and resource-based characteristics of the more marginal reaches of the state. For instance, the 1951 application for 115 new dispensaries was proposed along the lines of a five-mile radius plan, in which it was hoped that each 'place' in the UP would have a dispensary no further than five miles from it. In order to qualify, the government asked each applicant to provide detailed information about the population there (ranging from the overall number to the breakdown of the population by religion and caste, so that suitable practitioners could be sent to practice a suitable system for the location). Applicants also needed to supply a rundown of the variety of resources – government or informal – that already addressed the medical needs of citizens in the region, along with a survey of the land and building availability, all of which needed to be certified by the district dispensary.[42] In essence, these applications served as a way to ascertain the specific dimensions of health care through a free, localized assessment of resources and demands of the population therein. The dispensary also served as a representative of the state in areas where its reach was limited, acting as a capillary of state power in the rural stretches of Uttar Pradesh.

## Educating the Vaid

The most concentrated questions around planning a future for the indigenous medical systems were best articulated in questions about the future of their educational institutions. The terms for their existence had been decided in the 1920s and 1930s, as we saw in Chapter 3, while the dyarchic measures of government still allowed for a certain amount of flexibility and creativity in planning, and a great degree of innovation in thinking through the adaptation of traditional practices into modern ones. Under the Congress government of the post-1935 period, the planning focus had been on rural development, most notably addressed through a thorough mapping of the United Provinces into manageable land parcels serviced by registered practitioners. This variety of planning had also dominated the early postcolonial period through conversations about dispensaries, and by superficial forays into the question of Ayurveda and the public health.

The nature of an indigenous medical education had been cordoned off from government discussions in the 1930s and deemed the territory of vaids and hakims themselves. The Ayurvedic and Yunani Reorganization Committee, however, had recommended that this variety of education be streamlined into the overarching provincial plan for the province. The committee began by claiming that medical education was the sole responsibility of the state world over, as witnessed even by Uttar Pradesh's commitment to allopathic education.[43] The committee believed that it was therefore the province's responsibility to take account of the state of the educational institutions in which the Indigenous Medical Systems were taught and to ensure that they met the standards set out by government for medical education. The committee suggested a department-sponsored survey that would investigate each and every Board-affiliated college in order to ensure their quality.

The first set of concerns addressed the variety of teaching undertaken at the college. The first concern was over the questions of the practical application of skills adopted through direct patient contact, and was conversely a question about the role of the student in the community. The government was very concerned that each student should be assigned a bed in hospitals designated for their training, and suggested that the number of beds be increased at said hospitals to match the number of students in the upper-level classes. Moreover, a requirement was made that any new colleges wishing to open must be able to guarantee at least twice the number of beds as students in the first class before being given permission to affiliate with the board. Along with the acceptable number of beds followed a conversation about the adequate number of staff, which the board agreed should be no lower than 13 or 14 staff per college. Finally, where these or other material standards could not be met by the college, the government would undertake half of the expense of bringing the college up to par.[44] Colleges that could not improve would lose their affiliation and become dispensaries instead.

Ultimately, these concerns resulted not only in the streamlining of the schools but also in the bodies appointed to govern them, and to ascertain and maintain the quality of education. Having undertaken the survey, the government was able to identify changes that the board needed to make in order to properly regulate the schools, including the development of a singular granting system, a singular standard for examination (including a universal exam) and a standard set of classes necessary for certification.[45] In particular, the fear of deviation between colleges and programmes led the committee to surmise that the whole educational system was to be brought down by the perception of low standards. The committee questioned how it could ensure the quality of vaidyas and hakims if testing standards were not uniform, and recommended instead that tests be solely administered by the Faculty of Ayurveda affiliated with the Board of Indian Medicine. This latter point brought the debate around to its new permeation: whereas the substantial content of Ayurveda and later the practitioner had both been held up earlier as the reason for its failure to enter into modernity – and had both been subject to reform – its educational system was now identified as the major barrier to its scientific validity.[46]

The natural point of convergence of these suggestions made evident the gap left by the absence of a centralized college for the study of Ayurveda and Unani medicine that was conceptualized and run solely by the government. This institution would focus on the compilation, translation and updating of ancient texts of Ayurveda and Unani Tibb, and would set the standard for the rest of the province, while also serving as the regulatory mechanism that would set the syllabi, examination and certification process for other colleges and institutions. This institution would translate Ayurvedic and Tibbi texts from the Sanskrit and foreign language material into a set of Hindi-language textbooks that could then be supplied to colleges in UP and elsewhere in the subcontinent, or, indeed, the world.[47]

The descriptions of an educational system contained within this report mirrors the Chopra Report's suggestion of synthesis, but ultimately suggests a focus on Ayurveda that would have left little room for both Western medicine and other indigenous systems. The omission of a synthesis with Western medicine was an understandable outcome of curriculum planning, having been a rather lofty suggestion to begin with. However, the consistent deployment of Unani Tibb without any real commitment to the intellectual traditions behind it revealed an ongoing shift that positioned Ayurveda as the authentic indigenous tradition, and Unani Tibb as a vaguely foreign, newer other. The health apparatus within the province privileged Ayurveda at every turn, be it with the adoption of *Swasthavritta* (a distinctly Ayurvedic/Sanskritic principle), the opening of exponentially more schools and dispensaries under an Ayurvedic/Hindu rubric and a constant willingness to let Unani slide from government agendas when budgets and schemas needed editing.

The independent government's disavowal of Unani Tibb reflected a more abstracted but equally entrenched position on the Indian body. As we saw earlier, the modern resurrection of Ayurveda into a cogently political system had often relied upon the construction of certain 'truths' about the constitution of the Indian nation writ large, as well as the pragmatics of the populations in need of management. The indigenous bodies in question were consistently conceptualized as Hindu and casted, with a careful analysis of class and locality that situated citizens on the right or the wrong side of the 'healthy'. What emerges in these discussions of the Indigenous Systems of Medicine and the public health is the further marginalization of Muslim bodies and Islamic medicine from the spheres of health care that were meant to address the stretches of the nation. While Unani/Muslims existed in small part as targets for development, their gradual marginalization mirrored their shaky claim to belonging within the imagined nation of embodied Indians.

## Ayurveda diffused: Food as the newest health frontier

Where the context of state planning offers an image of progressive projections followed by pragmatic recoil, the public sphere again offers a counter-discourse of techno-science at the intersection of indigeneity that further shaped the plight of 'indigenous medicine' in this period. The Bhore and Chopra reports had ushered in an era where indigenous medicine had to live up to the rigours of scientific research, a challenge extrapolated more broadly into a revamping of tradition to fit the politics of the future. The idea of 'the future' introduced new terrains of possibility that had the result of expanding the paths towards – and indicators of – progress. The public heath focus on preventative medicine, translated provincially using the concept of *Swasthavritta*, set a precedent by linking 'modern' goals to 'ancient' practice, while positioning both within the spectre of successful planning for independence.

The interlinking of indigeneity, techno-science and prevention within imaginings of the future were by no means relegated to the realm of state planning. These messages percolated strongly in the public sphere, eventually coming to occupy the space carved out by Ayurveda and its related technologies earlier on, as seen in Chapter 3. These themes that had earlier on been focused around the constitution of the nation were now somewhat obfuscated so as to focus on the nation's literal constitution, starting from the ground up. Where the conclusions drawn in the 1930s were somewhat abstracted so as to fit themselves into the broader rubric of national belonging, the post-war/postcolonial context of advertising did not demand an adherence to this variety of nation-building. Instead, the focus in writing and advertising turned to the pragmatics of executing health, specifically on the exact conditions of healthy living.

At the beginning of the long journey to health was the question of diet, which manifested in the public sphere in a focus on food. Ranging from its preparation, to the health benefits of food, to new technologies for purchase, diet and nutrition came to dominate conversation about the body. The sale of diet, both as familiar products or new technologies, however, was difficult as it required a market literacy that was only beginning to emerge in the consumer markets of UP in this period. It also happened later than the other phenomena, with food advertisements emerging in the late 1930s, but dominating by the late 1940s and early 1950s. These advertisements attempted to sell both the brand and the technology, but they also had to convince the consumer that they wanted to revamp the feeding of their household. The sale of medicines to deal with intimate health issues had to overcome the informal, indigenous knowledge of the body usually held by certain members of the family or the community. However, there was a precedent for the purchase of medical products from outside sources. The introduction, branding and marketing of food products that already circulated widely in different contexts posed a particularly vexing problem to advertisers: food needed to be marketed in such a way as to convince the consumer to negotiate his or her relationship to its preparation. Ostensibly, consumers had to be taught to purchase ready-made goods, like curry pastes, tea bags and packaged spice mixes, which they were used to preparing for themselves. Furthermore, they also had to be convinced to pay the requisite mark-up.

The marketing of food products in this period is a very interesting if unexplored area of historical investigation. For instance, cookbooks were one of the products of both the new domesticity and the new-found popularity of printing, and revealed in their production, marketing and consumption a resonance with other cultural and social issues of importance. Arjun Appadurai has analysed the role of the cookbook industry in the larger project of constructing the national imaginary, emphasizing the consumption of food guides and cookbooks by the middle-class, female population, and the regulatory process of cooking, budgeting food allowances and planning meals suggested by the cookbook. Though his study addressed the 1970s, his model works well for the interwar years. One cookbook, compiled by J.A. Sarma in 1933, had ties with regional and religious groups. Though it was published by Chand, *Paka Vijnana* was commissioned by the *Adarsh-Parivar* ('Ideal Family'), a women's group affiliated with the newly founded *Hindi Sahitya Samellan* (the Hindi Academy), which encouraged the promotion of printed materials in Hindi, and had close ties with the Arya Samaj.[48] Included on the list of contributors were prominent Ranis and Maharanis. This book is part of the long tradition of books sold by women involved with charity work featuring recipes from individuals celebrated in other fields.[49] Another prominent cookbook in Allahabad, starting with a print run of 3,000 copies, had more overt political affiliations. Sri Mataprasada Gupta, a shop and restaurant owner, advertised in his cookbook items available for sale at his shop, and hence reinforced the relationship between the user of the book and other processes of consumption; he reminded readers that every cook is also a consumer, and should be targeted as such.[50] Yashoda Devi had also begun her career in medical publishing with a series of *Pak-Shastras* that explored the medical virtues of certain sorts of eating.[51]

The idea of a cookbook marketed to middle-class Indian women able to consume printed media raises important questions about the distribution of labour in the household and the constitution of 'women's work'. What was the relationship between the reader and the process of food preparation and distribution in the household? Missing from most guides was any discussion of which members of the household completed which aspects of food preparation, a time-consuming activity for most recipes. In particular, the absence of any mention of the role of the domestic servant stands out, which ignores the lived domestic experience of most of the women in a position to both purchase and read a cookbook. Food and its preparation has traditionally had great significance in Indian culture, as in most areas of the world, and the delineation of hierarchy and status in a household was often represented – and perhaps even reinforced – by the preparation of food.

The introduction of new food products, therefore, brought up many questions about the household, and about the role of the consumer within it. One of the most interesting themes in these ads was the negotiation of gender. The consumption of domestic goods was conceptualized in most guides of the period as the domain of the women of the household, along with the staff they controlled. Cookbooks were written with women in mind, characterized by the gendering of words in the feminine, and also in terms of the spaces and domains of power invoked in their wording. However, the figures in the advertisement were often male, and referenced a masculine figure of authority when describing the importance of adopting a particular substance. The Dalda Company's ad for a *vanaspati ghee* (hydrogenated vegetable oil used in place of clarified butter) to be used as the base of *sabzi* (vegetable) dishes provides a nice example of the genre.[52] The advertisement bears pictures of fresh fruit and vegetables laid out on a very basic-looking *thali*. Somewhat obfuscated by the image of fresh produce is a cookbook that bears the name of the company, and features a tin of the paste; a tiny image of a woman, her tiny arms spread over the large corner of the book, strives to open its pages. Atop this image is a slogan in large, bold print reading 'It will build strength' (*unko shaktidiya banaei*). Also featured across the text of the advertisement is the figure of a woman in a sari, clearly the mother figure, looking down at a group of boys running towards her; placed across from her is a man, legs spread, hands firmly on hips and chin facing upwards. Clearly he is meant to be the father, and the instigator of the family's shift in consumption. Lower down the page, the toothy grin of a boy is featured, with a cartoon bubble in which he informs the reader that 'father says that this will invigorate me'. The slogan of the company, found at the bottom of the advertisement, placed next to the company's name, reads 'for invigoration'.

It was the continued focus on the indigenous body in its 'natural' social and moral states that prevailed in this discourse: the healthy Indian family, the strong Indian boy child, the appropriate staging of marital intimacy as primarily focused on the offspring. While furthest from the realm of Ayurvedic practice and knowledge, there is a clear continuity between this sort of advertising and the most sophisticated Ayurvedic writing in the period, linked through the centrality of the body and its maintenance to the continued success of a nationalist politic. The themes implicit in them were the stock and trade of medical discourse across genres, reiterated here through the further positioning of domestic practices at the heart of all nationalist activity. The messages conveyed by the food advertisements are jarringly overt: men possess the knowledge of how to keep their family healthy, and women must be taught how to enact it; healthy, indigenous food is the key to building a strong family; a strong family begins with the health of its (boy) children. This strategy reveals the uncontested tenacity of these themes, which were secure enough in the realm of the social imaginary that they could be relied upon to introduce new behaviours into daily life.

Tea posed another interesting question for advertisers. Though tea had played a role in daily Indian life for a couple of centuries by this point, the mass consumption of tea is an inherently modern phenomenon. The infusion of herbs in water is an ancient practice, but the infusion of leaves resembling *Camellia Sensis* (the most common tea leaf), taken from the Indian *Sanjeevani* plant, did not become a popular practice until the mid-nineteenth century. The British attempted to cultivate tea from the 1860s on, resulting in the development of colonial tea plantations in the damp and elevated hills of various pockets of the subcontinent – including, for example, Darjeeling, Ootacamund, the Nilgiris, the lower Himalayan foothills and Munar in Kerala – as a way to curb the need to import tea from China.[53] Despite rough beginnings, South Asian (both from India and Ceylon) tea accounted for almost 90% of the tea consumed in the British Empire by the turn of the twentieth century.

Already a British leisure activity, running the gambit from genteel High Tea parties to the tea breaks (and suppers) of the working classes, the expansion of British tea production also called for the identification of new markets, and colonial contexts provided just that. Indians were taught to make tea a part of their everyday activities through a rigorous series of tea-drinking campaigns, beginning in the late nineteenth century, which often targeted key upper-caste/class families in regional districts, with the aim of having the trend trickle down. At the same time, a *chai* break was readily incorporated into the culture of work amongst the labouring classes, with the hope that labourers would incorporate the practice into their home life. Tea also became a staple of migratory and travel experiences, sold on train platforms, while cries of *chai garam! garam chai!* ('tea that's hot! hot tea!') became an accepted part of the travel encounter.[54]

This is where the Bharat Chai company enters the picture: not as the means to introducing a new product, but rather as a company interested in reclaiming an older product, staking their own claim as an Indian tea producer of Indian tea. The challenge facing Bharat Chai was to sway the Indian masses away from their affinity for tea as a British commodity, and to appreciate it as an Indian one instead.

A popular ad for Bharat Chai (Indian tea) proclaimed that tea was 'A Good Habit to Exert', but rather than employing a visual aide that privileged the genteel or practical aspects of drinking tea, the ad used a photograph of a young mother feeding her toddler tea.[55] Rather than using a teacup, the girl was learning to drink tea poured into a saucer – a common practice for cooling down hot tea and other hot substances in the subcontinent. Both mother and child are wearing *tikka* marks (noting their Hindu identity) and are adorned in simple but elegant jewellery, marking them as members of the middle class, and therefore reflecting the appropriate consumers of the substance and appropriate advocates for the practice. The intimate depiction of the act of feeding and being fed, in which both figures are smiling happily, lends to the framing of this activity as natural, comforting and safe.

This narrative of progress through the technologies of health was epitomized in a series of books that described the health and healing benefits of whole food products. Published by Dehota Pustak Bhandar in Delhi in the early 1950s, these cheap, 30-page booklets retailed for approximately Rs 11. They differed from other booklets in that they posted shiny, cardboard covers with illustrations of the fruit in question, lending an aesthetically pleasing air of value to them that other guides lacked. The author of the text was one Hakim Maulvi

Muhammad Abdullah Saheb, whose background was not discussed, but which leant the books a tie-in to an intuitive logic of indigeneity evoked with the reference to Unani professionalism. Each text was set up in a similar way, espousing the virtuous properties of different foods and plants, ranging from expected and popular household items like mangoes and pomegranates to the more unexpected species, namely the banyan tree (whose health benefits were transmitted while sitting under it in contemplation).[56] The benefits to the circulation of the blood, the illnesses of the eyes, the disturbances posed by fevers and colds, and many other conditions of dis-ease were addressed by these texts, along with specific directions about the preparation of the food product in question into a balm for various ailments.

What's striking about these texts is the shift within them away from the moral language of earlier texts, which couched similar information within the context of its ancient authenticity, to an aggressively modern rhetoric of scientific specificity. Gone was any whiff of a biomoral project, and in its stead came a very matter-of-fact deconstruction of the food or plant species in question, which began with a breakdown of the product at hand, ran pragmatically through its application in various contexts and ended with suggestions for its preparation and preservation in aid of the maintenance of household health.

Food writing ultimately revealed the tenacity of the indigeneity/technology/progress linkage, which was strong enough in the realm of the social imaginary that it could be relied upon to introduce new behaviours into daily life. Food writing and advertising were made readable and relevant by the earlier cementing of food as a marker, a determinant and a protector of health and wellness in the family, and subsequently the nation.

\*   \*   \*   \*

Food advertising, ultimately, did away with the figure of the author as mediator. There was no service offered in this instance, no expert advice passed on and no figure that could mediate the production of knowledge. Food advertising instead communicated the message to the public that they already were well aware of the choice they had to make, and casually informed them that the product would be available when they were ready to make the right choice. That it turned this argument on its head, using the health of the nation to introduce new food technologies further intertwined the steady relationship between the body, its needs and its place in the future of the Indian nation and state.

As we saw earlier, this era of experimental planning on a local, national and global level set the stage for an intimate familiarity with Ayurveda itself to become a designation of an attractively authentic and thoroughly modern understanding of the postcolonial Indian condition. Regardless of the stakeholder, who ranged from federal politicians to state bureaucrats to educational instructors to lay authors, all scrambled to lay claim to an authoritative knowledge of Ayurveda, and to formulate a politics of the possible for Ayurvedic medicine. As health governance became primarily concerned with the public health, Ayurveda and other appropriate questions of indigeneity rose to fulfil the gap left by the limits of allopathy. What emerged from these myriad attempts to use an Ayurvedic balm for the problems of the future was the fundamental nature of its transition: similar to all stakeholders was the understanding that the system to which they were laying claim had succeeded in becoming sufficiently techno-scientific so as to sustain its position as a cornerstone of state-building, and also of nation-building.

# Conclusion: Ayurveda's Indian Modernities

It is tempting to end this monograph by asserting that the government intervention into the state of Ayurveda, particularly as represented by the recasting of the vaid as political actor and the absorption of Ayurveda into the biopolitical practices of the state, resolved the tension between the poles of traditional medicine and medical modernities. Theoretically, it did: in the nineteenth century, Ayurvedic practitioners were deemed responsible for the sorry state of their tradition; but from 1919, Ayurveda was modernized and made relevant to the functioning of the state through the professionalization of the vaid. Vaids did the social and cultural work necessary in the Hindi public sphere to gain legitimacy as the new arbiters of tradition. Medical policy and medical institutions were developed to focus on their professional development as doctors, as legislators and as men of science. At the same time, the creativity around planning that was possible in the dyarchic moment was quickly rigidified into a more formalized biopolitics of late colonialism by the mid-1930s. These post-dyarchic institutions were anchored by the politics of health in the era of proto- and actualized Congress governance. The modern history of Ayurveda, it would seem, is characterized by the systematization of the practices grouped around the term into a cohesive, uncontested and operational rendering of the tradition in a form compatible with the normative politics of health governance.

Yet, a different history resonates as well. Colonial logic regarding Ayurveda championed the ancient, timeless and fallen tradition that Orientalists encountered in Sanskrit scriptures. Central to this conception of Ayurveda was a model of a globally resonant, culturally specific Ayurvedic past that was not reflected in the 'backward' medical practices and cultures encountered on the ground. The colonial state continually harkened back to the text even when faced with evidence of dynamic and varied medical cultures and practices that gave contours to the contemporary experience of Indians. When practitioners were given certain powers by the state, they worked steadily to reclaim textuality from the pandits, framing Ayurveda within the models of modernity that permeated. The authority of the text in social conceptions of Ayurveda – as cultural artefact, as instructive guide to ritual, as marker of professionalism – was not something that government policy could undo.

A genealogy of Ayurvedic development in the modern period also reveals the way in which Ayurveda went from being a composite of ideas about the body and its functioning to having a relatively singular meaning in the debates about embodied modernity in colonial North India. While there were certainly different 'Ayurvedas' at work, on all fronts (be they textual or pragmatic), it had to conform to the new cultures of health introduced by the colonial state. In turn, Ayurveda was recast across the board as an authentically indigenous system that could hold its own in a biomedical world. In the Hindi public sphere, middle-class ideals about nation, community and identity structured the image of Ayurveda presented. In the eyes of the provincial government, Ayurvedic institutions could provide key social interventions loosely framed as 'development' initiatives. Set against the backdrop of anti-colonial and communal politics, Ayurveda provided a traditional cure for the very modern ills of late colonial life.

In the previous chapters, we've seen how Ayurveda's genealogical transition through systematization began with the citational practices of early Orientalists looking for the textual impetus that could fit neatly into a global story of the evolution of medical knowledge and systems. This engagement was quickly silenced by the colonial state when the cosmopolitanism of liberal learning structures, embodied in the Native Medical College's hybrid curriculum, became too threatening. The imposition of a singular framework of power could not risk the challenge that hybridity posed to colonial dominance. The nineteenth-century engagement with the pragmatics of indigenous medical practice resituated the authority of the tradition with the body of the practitioner through the cooptation of his authority in attempts to carry out epidemic and pandemic health responses. The logic behind indigenous medical practices was not invoked in any formal way until the mid-1910s, at which point the state was petitioned by European supporters to investigate its propensity for health governance; the state in turn further marginalized its presence and attempted to strengthen that of biomedicine by inaugurating a set of medical acts that curtailed the practice of non-allopathic medicine within the institutional realm of the state.

In essence, this represented the first experience of colonial modernity, characterized by the Chatterjeean rule of colonial difference, in which racial ideologies of embodiment intersected with the politics of hegemonic governance during the 'high noon' of colonial rule. At the same time the confluence of imperial wealth, global positioning and hegemonic claims on culture during the period in which the sun never set on Empire bolstered the dynamic of imperial control and the beginnings of native resistance to it. This epoch of Empire in India came to a close following the devastations of the First World War, which resulted in the devastation of colonial economies, a serious blow to imperial mastery and resultant set of deferred promises for new arrangements of colonial governance to allow for steps towards semi-autonomous rule. These promises in action, as evidenced by the grand-scale political failure of dyarchy, only served to reveal the illiberal nature of British colonial rule, now characterized by the states of exception invoked and arranged to extinguish resistance and rebellion. The meta-narrative of late colonialism reveals the excesses of state violence and the configurations of Indian nationalism developed to challenge the colonial regime. This has resulted in the creation of paradigms of communalism, secularism, nationalism and radicalism that have been invoked to excavate the process of nation-building from the overly broad spectrum of the anti-colonial nationalist struggle. However, a different reading of interwar period reveals the shifting terrain of politics in the enclaves of Empire, namely the British provinces where the pragmatics of dyarchic and, later, semi-autonomous rule was transformed dramatically. The politics of the everyday became the domain over which provincial, municipal and other localized structures of diffused governance could exert power, in an exercise of governmentality that was at times mimetic and at times divergent from the earlier and contemporary practices of colonial governmentality.

This era of modernity proved a crucial one for the systematization of Ayurveda, for it was through the newly assigned provincial responsibility for the health of its citizens that the indigenous medical systems were resurrected and restructured into formalized systems capable of addressing and offering salve for the ills of modern subjects. In this encounter, Ayurveda was formally subject to the law of the state and incorporated into the growing bureaucracy as a way to further legislate and secure the health of the population. It was also in this context that Ayurveda was formally incorporated as a strategy of control and discipline, used to survey, organize and manage the population of the United Provinces. As the realm of formal politics became more and more provincialized, the spectre of governance transitioned from one in which the colonial state dictated the terms of politics to one in which a plurality of newly politicized entities sought to exercise power, resulting in a plurality of governmentalities on the ground. Central to these machinations of power around questions of citizenship and belonging were the looming problems of the division of assets and resources amongst the 'teeming masses', themselves divided along communal and otherly identitarian lines. Ayurveda, in essence, was further modernized around this concept of its role in health governance.

Ayurveda's final encounter with the shifting terrains of Indian modernity was made operational through the shift in politics of the Second World War. Where the interwar period was pre-occupied with both the anti-colonial struggle and the responsibilities of governance on the ground, the post-war period initiated an era of optimism wrapped up in the promise of a postcolonial future. The looming promise of independence, followed by the violent enactments of its inception, resulted in a reorientation of health governance to the goals of development planning for a successful future. While these strands of practice were characteristic of Congress governance from the late 1930s on, the 1940s saw the mass adoption of a techno-scientific future for the nation. Centralized agencies adopted scientific methodologies for measuring questions of health, disease, prevention and cure for the Indian masses, making proposals for strategies too expansive and expensive to employ. The Indigenous Systems of Medicine, having been made modern enough to help govern decades earlier, were invoked as immediate salves to the postcolonial problematic of the tenuous public health (or entire lack thereof). At the provincial level, Ayurveda was specifically invoked to lend a traditionally authentic Hindu idiom to the new machinations of the public health apparatus. Ultimately, by the early 1950s, the biomoral and biopolitical converged in the form of an aggressive Ayurvedic system of medicine, capable of ensuring the public health, while simultaneously policing the bounds of the public through an insistence upon Hindu cultural belonging.

# Notes

## Introduction: Ayurveda in Motion

1. The author must admit at this juncture to being fully susceptible to the charm (both olfactorial and visual) of these products, along with the lure of self-indulgence that only a beautiful bottle of lotion can induce. I am the owner of a wide range of these products, both bought locally in Canada and during periods of fieldwork in India. The ploys of neo-liberal marketing strategies have not impeded my ability to identify and critique the greater systems at work in constructing this experience of consumption. I am also not alone in my experience as participant-observer of Ayurvedic luxury in motion; the anthropologist Jean Langford begins and ends her excellent book on Ayurveda with a recounting of an Ayurvedic massage she received in Kerala. See Jean Langford, *Fluent Bodies: Ayurvedic Remedies for Postcolonial Imbalance* (Durham: Duke University Press, 2002).
2. In the fall of 2012, for instance, a 200 ml bottle of Kama brand almond oil cost Rs.1075 at the Kama store in Khan Market. A bottle of Khadi almond oil, here appealing to a different affective ideal, bought in a pharmacy around the other side of the market, cost Rs 160 for 150 ml.
3. See Mark Harrison, *Public Health in British India: Anglo-Indian Preventive Medicine, 1859–1914* (Cambridge: 1994), p. 227. Also see Dominik Wujastyk, *The Roots of Ayurveda*, 3rd edn (London: 2003), p. 4.
4. These approaches are best encapsulated in the works of Guy Attewell, *Refiguring Unani Tibb* (Hyderabad: Orient Longman, 2007); Projit Bihari Mukharji, *Nationalizing the Body: The Medical Market, Print and Daktari Medicine* (Anthem Press, 2009); Madhuri Sharma, *Indigenous and Western Medicine in Colonial India* (Delhi: Cambridge University Press, 2012); Kavita Sivaramakrishnan, *Old Potions, New Bottles: Recasting Indigenous Medicine in Colonial Punjab* (Hyderabad: Orient Longman, 2006).
5. For instance, Sheldon Pollock's work on the 'death' of Sanskrit in the age of communal politics attempts to historicize the meaning of language. See Sheldon Pollock, *The Language of the Gods in the World of Men: Sanskrit, Culture, and Power in Premodern India* (Berkeley: University of California Press, 2006). Michael Dodson's groundbreaking work on the re-appropriation of Sanskrit as a language of scientific education by pandits reconfiguring their authority in colonial India makes a similar intervention. Please see Michael S. Dodson, *Orientalism, Empire, and National Culture: India, 1770–1880* (Basingstoke: Palgrave Macmillan, 2007).
6. Dominik Wujastyk, one of the most eminent scholars of ancient medical knowledge reflected in Sanskrit texts (and the author of the most thorough guide to extant medical texts of Ayurvedic medicine), has begun to take this approach on board as a mode of investigation and has produced the most thorough studies of Ayurveda in context. See, for instance, Dominik

Wujastyk, *The Roots of Ayurveda*, 3rd edn (London: Penguin, 2003). More recently Dagmar Wujastyk has reflected upon Ayurvedic logics within the context of the evolving field of bioethics in modernity; please see Dagmar Wujastyk, *Well-Mannered Medicine: Medical Ethics and Etiquette in Classical Ayurveda* (Oxford: Oxford University Press, 2012).

7. See R. Ramasubban, *Public Health and Medical Research in India* (Stockholm: Swedish Agency for Research Cooperation with Developing Countries, 1982); R. Ramasubban, 'Imperial Health in British India 1857–1900', in *Disease, Medicine & Empire: Perspectives on the Medicine and the Experience of European Expansion*, ed. R.M. Macleod and M. Lewis (London: Routledge, 1988).

8. David Arnold, *Colonizing the Body: State Medicine and Epidemic Disease in Nineteenth-Century India* (Berkeley: University of California Press, 1993), p. 12–13.

9. M. Harrison, *Public Health in British India: Anglo-Indian Preventive Medicine 1859–1914* (Cambridge: Cambridge University Press, 1994), p. 226–227.

10. Ishita Pande, *Medicine, Race and Liberalism in British Bengal: Symptoms of Empire* (London: Routledge, 2009). The broader topic of liberalism has become a recent theme amongst Indian historians who frame liberalism as an active project as well as a historical ideology. Please see Shruti Kapila, *An Intellectual History for India* (Cambridge: Cambridge University Press, 2011).

11. Kavita Sivaramakrishnan, Old Potions, New Bottles: Recasting Indigenous Medicine in Colonial Punjab (Hyderabad: Orient Longman, 2006).

12. Guy Attewell, *Refiguring Unani Tibb* (Hyderabad: Orient Longman, 2007).

13. Alavi's work remains an exception here as well, though the context she explores is the global Islamic intellectual world, rather than the particularity of empire in India.

14. See S. Marks, 'What Is Colonial About Colonial Medicine? And What Has Happened to Imperialism and Health?', *Social History of Medicine* 10, no. 02 (1997): 205–219. Waltraud Ernst has grappled with similar issues, albeit with very different conclusions in W. Ernst, 'Beyond East and West: From the History of Colonial Medicine to a Social History of Medicine (s) in South Asia', *Social History of Medicine* 20, no. 3 (2007): 505–524.

15. Warwick Anderson, 'Where Is the Postcolonial History of Medicine?', *Bulletin of the History of Medicine* 72, no. 3 (1998): 522–530.

16. Arnold, *Colonizing the Body*.

17. Arnold, *Colonizing the Body*, p. 8. Also see G. Burchell, C. Gordon, and P. Miller, *The Foucault Effect: Studies in Governmentality* (London: Harvester Wheatsheaf, 1991).

18. Sarah Hodges, 'Looting the Lock Hospital in Colonial Madras during the Famine Years of the 1870s'. *Social History of Medicine*, 18 (2005): 379–398, p. 380. Also see Mark Harrison, *Climates & Constitutions: Health, Race, Environment and British Imperialism in India, 1600–1850* (Delhi: 1999). Kavita Philip explores the applied science of race in other debates about nature, environment and resources in larger discourses of environmentalism. See Kavita Philip, *Civilizing Natures: Race, Resources, and Modernity in Colonial South India* (Camden: Rutgers University Press, 2003).

19. Of course, the idea that 'gender is a relation of power' predates Arnold and harkens back to Joan Scott's seminal works in defining gender as a category of historical analysis; see J.W. Scott, *Gender and the Politics of History* (New York: Columbia University Press, 1989). However, Arnold's model has proved useful in the following works: M. Lal, 'The Ignorance of Women Is the House of Illness: Gender, Nationalism, and Health Reform in Colonial North India', in *Medicine and Colonialism: The Politics of Identity*, ed. M.P. Sutphen (London, 2003); D.M. Peers, 'Privates Off Parade: Regiments and Sexuality in the Nineteenth Century Indian Empire', *International History Review* 20, no. 4 (1998): 823–854; P. Levine, *Prostitution, Race and Politics: Policing Venereal Disease in the British Empire* (London: Routledge, 2003); Sarah Hodges, 'Looting the Lock Hospital in Colonial Madras During the Famine Years of the 1870s', *Social History of Medicine* 18, no. 3 (2005): 379–398.

20. See Sarah Hodges, *Contraception, Colonialism and Commerce: Birth Control in South India, 1920–1940* (Aldershot: Ashgate, 2008).

21. See Stephen Legg, *Spaces of Colonialism: Delhi's Urban Governmentalities* (Oxford: Wiley-Blackwell, 2007), introduction.

22. This argument has been most notably forwarded in Ann Laura Stoler, *Race and the Education of Desire: Foucault's History of Sexuality and the Colonial Order of Things* (Durham: Duke University Press, 1995). See also Anderson, 'Where Is the Postcolonial History of Medicine?'; Hodges, 'Looting the Lock Hospital in Colonial Madras During the Famine Years of the 1870s'; M. Vaughan, *Curing Their Ills: Colonial Power and African Illness* (Cambridge: Cambridge University Press, 1991).

23. The Althusserian roots of Foucauldian theory work to inflect his work with ruminations on class, even if indirectly. That said, the conception of class at work in Europe is only somewhat appropriate for the semi-industrial and overwhelmingly rural (and occasionally feudal) state of the interwar economy and its related socio-economic structures.

24. In so doing, the question of population foments the three forms of possibilities that the *dispositif* or apparatus holds in Foucauldian thought: the heterogeneous ensemble, the connective forces, the urgent need. Please see M. Foucault, 'The Confession of the Flesh', *Power/knowledge: Selected Interviews and Other Writings* 1977 (1972): 194–228.

25. These now famous debates characterize the intellectual tensions of the 1990s. The legacy of 'schools' and their influence on eighteenth-century history is well-summarized in R. Travers, 'The Eighteenth Century in Indian History: A Review Essay', *Eighteenth-Century Studies* 40, no. 3 (2007): 492–508.

26. This move away from nationalist historiography towards a more fractured one focused on elite constructions and deployments of nationalism, towards a view of history from its non-elite and, ultimately, subaltern subjects has been well documented within reflections upon the Subaltern Studies School. Framed first as a Marxist (and somewhat Annaleist) critique of elite-focused histories, this movement elaborated the critiques of French cultural theory – most poignantly, Foucauldian readings of power (filtered through the work of Edward Said) and the Derridian method of deconstruction – to think through the historical subjectivity of those actors not documented in the normative record, in essence producing a sophisticated Indian history from below. In the proceeding decades, scholars have heatedly debated

the techniques and ideological pinnings of the Subaltern Studies School; at this moment, more than two decades after its introduction, the perspectives offered by the cultural turn have become historiographical commonplace and the debates themselves have become a vital part of the historiography of modern India. For a recounting of the initial debates (and their legacy), please see Gyan Prakash, 'Subaltern Studies as Postcolonial Criticism', *The American Historical Review* 99, no. 5 (1994): 1475–1490; Rosalind O'Hanlon, 'Recovering the Subject: Subaltern Studies and Histories of Resistance in Colonial South Asia', *Modern Asian Studies* 22, no. 1 (1988): 189–224; D. Chakrabarty, *Habitations of Modernity: Essays in the Wake of Subaltern Studies* (University of Chicago Press, 2002). Vinay Lal has elegantly reflected upon the influence of the Subaltern Studies turn in the decades following its introduction; see V. Lal, 'Subaltern Studies and Its Critics: Debates over Indian History', *History and Theory* 40, no. 1 (2001): 135–148.

27.  Gyanendra Pandey, *The Construction of Communalism in Colonial North India* (Delhi: Oxford University Press, 2006).

28.  Here I am borrowing once again from Stephen Legg.

29.  This line of argument is actually crucial to the 'Cambridge School' approach to the history of nationalism, in which historians argued that the local elites simply took over the vestiges of power on the ground in local contexts, rendering the structures of nationalism in interwar India massively elitist ones. The texts that most rigorously delineate this perspective remain A. Seal, *The Emergence of Indian Nationalism: Competition and Collaboration in the Later Nineteenth Century*, vol. 1 (Cambridge: Cambridge University Press, 1971); J. Gallagher, G. Johnson, and A. Seal, *Locality, Province and Nation: Essays on Indian Politics, 1870 to 1940* (London: CUP, 1973).

30.  The most effective work on this topic is T.C. Sherman, *State Violence and Punishment in India* (London: Routledge, 2009).

31.  This latter category of internationalism is a topic of great importance to contemporary historians, as evidenced in recent works, including S. Bose, *His Majesty's Opponent: Subhas Chandra Bose and India's Struggle Against Empire* (Cambridge: Harvard University Press, 2011); Michele L Louro, 'Rethinking Nehru's Internationalism: The League Against Imperialism and Anti-imperial Networks, 1927–1936', *Third Frame: Literature, Culture and Society* 2, no. 3 (September 2009): 79–94; Carolien Stolte, 'Enough of the Great Napoleons!' Raja Mahendra Pratap's Pan-Asian Projects (1929–1939)', *Modern Asian Studies* 46, no. Special Issue 02 (2012): 403–423; Sugata Bose and Kris Manjapra, eds., *Cosmopolitan Thought Zones: South Asia and the Global Circulation of Ideas* (London: Palgrave Macmillan, 2010).

32.  Sunil Amrith has delineated in several works the necessity of looking at India's engagement with internationalist political formations, like the League of Nations, which works to insert India into larger global stories of political development, while deparochializing the realm of politics at home. See especially S.S. Amrith, *Decolonizing International Health: India and Southeast Asia, 1930–1965* (Basingstoke: Palgrave Macmillan, 2006); Sunil S. Amrith, 'Food and Welfare in India, C. 1900?1950', *Comparative Studies in Society and History* 50, no. 04 (2008): 1010–1035.

33.  Mrinalini Sinha's pioneering work on the 'Mother India' scandal is perhaps most sensitively emblematic of this variety of transnational exchange with

regard to gender and sexuality. Please see M. Sinha, *Specters of Mother India: The Global Restructuring of an Empire* (Durham: Duke University Press, 2006).
34. F. Robinson, 'Nation Formation: The Brass Thesis and Muslim Separatism', *Journal of Commonwealth and Comparative Politics* XV (1977): 215–230; P.R. Brass, 'A Reply to Francis Robinson', *Journal of Commonwealth and Comparative Politics* XV (1977): 231–233; L. Brennan, 'The Illusion of Security: The Background to Muslim Separatism in the United Provinces', *Modern Asian Studies* 18, no. 2 (1984): 237–272; P.R. Brass, *Language, Religion and Politics in North India* (Cambridge: CUP, 1974). P.R. Brass, *Language, Religion and Politics in North India* (Cambridge: 1974); P.R. Brass, 'A Reply to Francis Robinson', *Journal of Commonwealth and Comparative Politics, XV* (1977), 231–233; Francis Robinson, 'Nation Formation: the Brass Thesis and Muslim Separatism', *Journal of Commonwealth and Comparative Politics, XV* (1977), 215–230. Lance Brennan, 'The Illusion of Security: The Background to Muslim Separatism in the United Provinces', *Modern Asian Studies,* 18 (1984), 237–272; Robinson, 'Nation Formation: the Brass Thesis and Muslim Separatism'.
35. W. Gould, *Hindu Nationalism and the Language of Politics in Late Colonial India* (Cambridge: Cambridge University Press, 2004).
36. F. Orsini, *The Hindi Public Sphere, 1920–1940: Language and Literature in the Age of Nationalism* (Delhi: Oxford University Press, 2002); V. Dalmia, *The Nationalization of Hindu Traditions: Bharatendu Harishchandra and Nineteenth Century Benaras* (Delhi: Oxford University Press, 1997); A. Rai, *Hindi Nationalism* (Delhi: Three Essays Collective, 2001).
37. Sanjay Joshi, *Fractured Modernity: Making of a Middle Class in Colonial North India* (Delhi: 2002), pp. 2–3. Additionally, William Gould's work on the language of nationalist discourse has linked popular ideas to the realm of high politics; see Gould, *Hindu Nationalism and the Language of Politics in Late Colonial India.*
38. Dalmia, *The Nationalization of Hindu Traditions: Bharatendu Harishchandra and Nineteenth Century Benaras.*
39. Orsini, *The Hindi Public Sphere, 1920–1940: Language and Literature in the Age of Nationalism.*
40. Rai, *Hindi Nationalism.*
41. Gupta, *Sexuality, Obscenity, Community: Women, Muslims, and the Hindu Public in Colonial India.*
42. This is a point Sanjay Joshi elaborates further in his study of the emergence of the middle class in the United Provinces. S. Joshi, *Fractured Modernity: Making of a Middle Class in Colonial North India* (Delhi: Oxford University Press, 2002).
43. The initial model offered is offered, of course in J. Habermas, *Structural Transformation of the Public Sphere: An Inquiry into a Category of Bourgeois Society* (Cambridge: MIT Press, 1991). Critics of Habermas have long since established the limits of civic participation, and have looked to counter-spheres for alternative modes and products of cultural production. Please see N. Fraser, 'What's Critical About Critical Theory? The Case of Habermas and Gender', *New German Critique* (1985): 97–131; N. Fraser and C. Calhoun, eds., *Habermas and the Public Sphere* (Cambridge: MIT Press, 1992); Michael Warner, *Publics and Counterpublics* (New York: Zone Books, 2002). In the

Indian context, the concept of the public sphere has been most successfully deployed (and critiqued by Francesca Orsini, whose work we will explore in Chapter 3). Please see Orsini, *The Hindi Public Sphere, 1920–1940: Language and Literature in the Age of Nationalism.*

44. See, for instance, Partha Chatterjee, *The Nation and Its Fragments: Colonial and Postcolonial Histories* (Princeton: Princeton University Press, 1993). Also see M. Goswami, *Producing India: From Colonial Economy to National Space* (Chicago: University of Chicago Press, 2004).

45. S. Beth, 'Hindi Dalit Autobiography: An Exploration of Identity', *Modern Asian Studies* 41, no. 03 (2007): 545–574; B. Narayan, 'Inventing Caste History: Dalit Mobilisation and Nationalist Past', *Contributions to Indian Sociology* 38, no. 1–2 (2004): 193–220; B. Narayan, 'Demarginalisation and History Dalit Re-Invention of the Past', *South Asia Research* 28, no. 2 (2008): 169–184.

46. Chatterjee, *The Nation and Its Fragments.*

# 1 Historicizing Ayurveda: Genealogies of the Biomoral

1. V.W. Karambelkar, *The Atharva-veda and the Ayur-veda* (Nagpur, 1961), 21–2.

2. It is important to note that the term 'Ayurveda' does not originate in the Atharvaveda, though the traditions related to it are found there. Ayurveda as catchall title is best dated to the emergence of the first *Caraka* and *Susruta samhitas,* where it was consciously employed to refer broadly to a set of traditions. See K.G. Zysk, *Medicine in the Veda: Religious Healing in the Veda with Translations from the Rg Veda and the Atharvaveda and Renderings from Corresponding Ritual Texts,* vol. 3 (Delhi: Motilal Banarsidass, 1996), introduction.

3. D.P. Chattopadhyaya, *History of Science and Technology in Ancient India* (Calcutta: Firma KLM, 1986), 32.

4. Dominik Wujastyk, *The Roots of Ayurveda,* 3rd edn (London: Penguin, 2003), xxix; K.G. Zysk, *Asceticism and Healing in Ancient India: Medicine in the Buddhist Monastery* (Oxford: Oxford University Press, 1991), 22–4.

5. Wujastyk, *The Roots of Ayurveda,* xxix.

6. A.L. Basham, *A Cultural History of India* (Oxford: OUP, 1975), 19–20.

7. Wujastyk, *The Roots of Ayurveda,* 5.

8. Ibid., 6. The actual question of influence, however, is not one that can be resolved, as the textual material that might trace this sort of trajectory is no longer extant. The similarities between the different traditions hint to the possibilities of exchange, but it is difficult to put forth a straightforward argument about exchange. I want to thank Dagmar Wujastyk for this particular insight into the question of influence.

9. Wujastyk, *The Roots of Ayurveda,* 63–4.

10. P.V. Sharma, *Susruta Samhita* (Varanasi: Chaukhamba Press, 1999).

11. Wujastyk, *The Roots of Ayurveda,* 63.

12. White identifies in this latter practice a connection between Ayurveda and yogic principles espoused similarly in ancient texts of philosophy. See D. White, *Kiss of the Yogini: 'Tantric Sex' in Its South Asian Contexts* (Chicago: Chicago University Press, 2003), 17. Dagmar Benner Wujastyk

also comments on notions of health and the quality of life that underlie these principles. See D. Benner, 'Traditional Indian Systems of Healing and Medicine: Ayurveda', in *Encyclopedia of Religion*, 2004, 5–6.

13. Sharma, *Susruta Samhita*, 155–7.
14. Ibid., 55.
15. Ibid., 157–57 & Chapter XV. Wujastyk also makes the argument that the view of the body in this conceptualization of cooking is unwaveringly male, and that the authors of these texts do not offer an alternative form of semen-production found in women's bodies. Wujastyk, *The Roots of Ayurveda*, xvii–xix.
16. Dagmar Benner Wujastyk explains it as such: 'The treatment of mala-related illnesses corresponds to the treatment of the dosas in that the malas increase or decrease, but also as the dosas are seen as the underlying cause of the increase or decrease of the malas.' Benner, 'Traditional Indian Systems of Healing and Medicine: Ayurveda', 7.
17. See, for instance, his pioneering work on the corporeality in South Asia; see D.G. White, *The Alchemical Body: Siddha Traditions in Medieval India* (Chicago: University of Chicago Press, 1996). His work on *tantra* makes equally salient points regarding the intersection of medicine and the body; in White, *Kiss of the Yogini: 'Tantric Sex' in Its South Asian Contexts*.
18. F. Zimmermann, *The Jungle and the Aroma of Meats*, 1982. Reference to the corruption of the original Sanskritic term is found on p. 2.
19. Ibid., 2–7 & chapter 1.
20. Ibid., 3.
21. Sharma, *Susruta Samhita*, 339.
22. Zimmermann, *The Jungle and the Aroma of Meats*, 3–4.
23. White, *The Alchemical Body: Siddha Traditions in Medieval India*, 24–5.
24. The question of an 'Aryan' invasion has been one of great historical and political import. Both the debates and thoughts towards a resolution of the historical questions around it can be found in R. Thapar, *The Aryan: Recasting Constructs* (Gurgaon: Three Essays Collective, 2008).
25. H.J.J. Winter, 'Science', in *A Cultural History of India*, ed. A.L. Basham, 2nd edn (Delhi: OUP, 1997), 147–8.
26. Zysk, *Asceticism and Healing in Ancient India: Medicine in the Buddhist Monastery*, 12. Zysk is referring to the dominant scholarship on the excavation of Mohenjo-Daro and the meaning of its findings, including works produced at the time of its discovery as well as more critical commentaries on the politics of the excavation.
27. Ibid., 13.
28. Ibid., 12.
29. R. Thapar, 'Renunciation: The Making of a Counter-culture?', in *Cultural Pasts: Essays in Early Indian History*, ed. R. Thapar (Delhi: Oxford University Press, 2000), 876–913. This is also the well-trod ground of White, *The Alchemical Body: Siddha Traditions in Medieval India*.
30. For an extensive history of medicine in the Buddhist traditions and communities of ancient India, see Zysk, *Asceticism and Healing in Ancient India: Medicine in the Buddhist Monastery*.
31. Ibid., 39–41.
32. Ibid., 43.

33. Ibid., 49 52. See also chapter 4, 'Indian Medicine in Buddhism Beyond India', for an in-depth comparison of Sanskrit, Pali, Tibetan and Chinese sources that give insight into a plethora of linguistic, medical and cultural linkages rooted in shared aspects of medical traditions.
34. See B. Pati, 'Negotiating with Dharma Pinnu: Towards a Social History of Smallpox in Colonial Orissa', *Canadian Bulletin Medical History*, 19, no. 2 (2002): 477–92.
35. See, for example, Madhu Ramnath's attempt to resurrect a truly 'indigenous' science in Pati Biswamoy, 'Negotiating with Dharma Pinnu: Towards a Social History of Smallpox in Colonial Orissa', *Canadian Bulletin Medical History*, 19 (2002): 477–92.
36. Zysk, *Asceticism and Healing in Ancient India: Medicine in the Buddhist Monastery*, 21–2.
37. Wujastyk, *The Roots of Ayurveda*, xvi–xvi, 10, 13.
38. See J. Legge, *A Record of Buddhistic Kingdoms Being an Account by the Chinese Monk of His Travels in India and Ceylon (AD 399–414) in Search of the Buddhist Books of Discipline* (Oxford: Clarendon Press, 1886).p. number
39. R. Thapar, *Early India: From the Origin to AD 1300* (London: Penguin, 2002), 258.
40. Ibid.
41. Thapar, 'Renunciation: The Making of a Counter-culture?'.
42. Ibid., 904.
43. W. Doniger and S. Kakar, *Vatsyayana's Kama Sutra* (Oxford: OUP, 2002), introduction.
44. The reference to Foucault is from Volume 1 of the History of Sexuality, wherein he describes the 'ars erotica' of the Asian world as a counter to the scientific function of equivalent texts in the Western world. See M. Foucault, *The History of Sexuality: Vol 1, The Will to Knowledge* (Harmondsworth: Penguin, 1990), 57. Doniger & Kakar cite it on p. xv of their text.
45. Doniger and Kakar, *Vatsyayana's Kama Sutra*, xx–xxi.
46. Ibid., 174. This is found in Book Seven of the Kamasutra, verse 1.4.6.
47. This is clear throughout David White's readings of Tantric practice, where the *Susruta* and *Caraka Samhitas* are continually invoked to give context for the substantive claims about the body made in Tantric texts. See White, *Kiss of the Yogini: 'Tantric Sex' in Its South Asian Contexts*, 39, 52, 224, 228.
48. In a rather interesting turn, Seema Alavi makes evident the presence of Hindustani medicine in Baghdad during the Abbasid caliphate, where Caliph Mansur hosted Hindustani doctors and had the *Sushruta* and *Sameeka Samhitas* translated. See Seema Alavi, 'Medical Culture in Transition: Mughal Gentleman Physician and the Native Doctor in Early Colonial India', *Modern Asian Studies*, 42, no. 5 (September 2008): 853–97.
49. Abu Fazl, *The Ain-i-Akbari by Abul Fazl-i-Allami*, trans. H.S. Jarret (Calcutta: Asiatic Society of Bengal, 1894), 224–5.
50. J. Alter et al., 'Heaps of Health, Metaphysical Fitness: Ayurveda and the Onthology of Good Health in Medical Anthropology (and Comments and Reply)', *Current Anthropology*, 40, no. Supplement: Special Issue: Culture, A Second Chance? (1999): S43–S66.
51. Ibid., S45.
52. Ibid.

53. See *Asian Medical Systems,* ed. by Charles Leslie (Berkeley: University of California, 1976), introduction. For Leslie's call to action, also see Leslie and Young's collection *Paths to Asian Medical Knowledge,* ed. Charles Leslie and Allan Young (Berkeley: University of California Press, 1992), introduction. This was an anthology in which a new generation of scholars (and, in some cases, not so new) of the social study of medicine revisited Leslie's initial project in light of advances in the field.
54. Leslie's contribution is elegantly summarized in David Arnold, 'Plurality and Transition: Knowledge Systems in Nineteenth Century India' (presented at The Princeton History of Science Seminar, Princeton, October 2003).
55. Jean Langford, *Fluent Bodies: Ayurvedic Remedies for Postcolonial Imbalance* (Durham: Duke University Press, 2002), 14.
56. Arnold, 'Plurality and Transition: Knowledge Systems in Nineteenth Century India'.
57. See P.B. Mukharji, *Nationalizing the Body: The Medical Market, Print and Daktari Medicine* (London: Anthem Press, 2009), 25–7.
58. Guy Attewell, *Refiguring Unani Tibb* (Hyderabad: Orient Longman, 2007), 24.
59. D. Varma, 'From Witchcraft to Allopathy: Uninterrupted Journey of Medical Science', *Economic and Political Weekly* (19 February 2006): 3605–11.
60. Though a theme in Marriott's writing from the late 1960s on, it is his collected anthology on the transactional categories of Hindu social organizations in which his work and that of his students was fleshed out. See M. Marriott and R. Inden, 'Towards an Ethnosociology of South Asian Caste Systems', in *The New Wind: Changing Identities in South Asia,* ed. K. David (The Hague: Mouton Publishers, 1977).
61. M. Marriott, *India Through Hindu Categories* (Delhi: Sage, 1990), 5.
62. Marriott and Inden, 'Towards an Ethnosociology of South Asian Caste Systems'.
63. See M. Marriott, 'Interpreting Indian Society: A Monistic Alternative to Dumont's Dualism', *The Journal of Asian Studies,* 36, no. 1 (1976): 111. The intervention that Marriott seeks to make is of course in direct response to Dumont's neat taxonomies of hierarchy, found in his employment of a 'dualistic' model for understanding the meaning of caste, and which provided the basis for his 'Homo Hierarchus'.
64. J. Parry, *Death in Banaras* (Cambridge: Cambridge University Press, 1994), 113.
65. Marriott and Inden, 'Towards an Ethnosociology of South Asian Caste Systems', 228.
66. Jonathan Parry, however, problematizes the supposed coherence of metaphysical identification that Marriott insists upon. While upholding Marriott's model as a viable critique of Dumont's dualism, and while employing the model of the substance-code connection therein as a foundation stone of his own investigation, Parry allows for 'a robust and stable sense of self' to which most of his participants gave voice despite their monistic, dividual connectedness. Parry suggests that the model that Marriott and colleagues have created is somewhat overdrawn, and, by way of intervention, poses this question to them: 'how indeed can anybody ever decide with whom, and on what terms, to interact?' See Parry, *Death in Banaras,* 114.

67. Please see Jacob Copeman, *Veins of Devotion: Blood Donation and Religious Experience in North India* (Rutgers University Press, 2009).

68. This is true across the board in studies of communication and culture in the South Asian context, and will be discussed in Chapter 4. See, for instance, F. Orsini, *The Hindi Public Sphere, 1920–1940: Language and Literature in the Age of Nationalism* (Delhi: Oxford University Press, 2002); C.A. Bayly, *Empire and Information: Intelligence Gathering and Social Communication in India, 1780–1870* (Cambridge: CUP, 1996).

69. Langford, *Fluent Bodies: Ayurvedic Remedies for Postcolonial Imbalance*, 14–15.

70. Ibid., 15. Langford in fact makes reference to Zimmermann's work in this context – the identification of *namamala* is his.

71. Please see J. Alter, *Gandhi's Body: Sex, Diet and the Politics of Nationalism* (Philadelphia: University of Pennsylvania Press, 2000).

72. Lawrence Cohen, 'The Other Kidney: Biopolitics Beyond Recognition', *Body & Society*, 7, no. 2–3 (2001): 15.

73. D.J. Haraway, *Simians, Cyborgs, and Women: The Reinvention of Nature*, 41 (Routledge, 1991), 162.

74. Cohen, 'The Other Kidney', 21.

75. Ibid., 22. See also G. Agamben and D. Heller-Roazen, *Homo Sacer: Sovereign Power and Bare Life* (Stanford: Stanford University Press, 1998).

76. The recent anthropological work of Jean Langford makes evident the investment in Ayurveda's situatedness outside of the perils of medical modernity. See Langford, *Fluent Bodies: Ayurvedic Remedies for Postcolonial Imbalance*.

77. Jones, 'On the Literature of the Hindus, from Sanscrit, communicated by Goverdhan Caul, with a short commentary', 340.

78. C.A. Bayly details the diverse make-up of these projects, filtered the aspiration to a cosmopolitan liberalism, in C.A. Bayly, 'Rammohan Roy and the Advent of Constitutional Liberalism in India, 1800–1830', *Modern Intellectual History*, 4, no. 01 (2007): 25–41. Also see C.A. Bayly, *Recovering Liberties: Indian Thought in the Age of Liberalism and Empire* (Cambridge: Cambridge University Press, 2012), introduction.

79. IOR F/4/737 no. 20085, 3.

80. IOR F/4/737 no. 20085, Medical Board to Military Dept Government of Bengal, (24 May 1822), 24.

81. Sir Thomas Babington Macaulay's Minute on Education banned funding to any educational institution that taught in vernacular languages, resulting in the transformation or closure of existing colleges – and, most notably for Ayurvedic education, the closure of the Native Medical School, and the Anglicisation of the Native Medical Hospital. Ishita Pande elegantly explores the ramifications of the legislation upon the field of medicine in Ishita Pande, *Medicine, Race and Liberalism in British Bengal: Symptoms of Empire* (London: Routledge, 2009), 65–7.

82. The event is discussed in David Arnold, *Colonizing the Body: State Medicine and Epidemic Disease in Nineteenth-Century India* (Berkeley: University of California Press, 1993), 6.

83. Seema Alavi has written extensively on the shift in Unani professional organization, and also on the advent of Unani medical journals as a popular genre of Urdu writing. See Seema Alavi, *Islam and Healing: Loss and Recovery of an Indo-Muslim Medical Tradition 1600–1900* (Delhi: Permanent Black, 2007).

84. Alavi, 'Medical Culture in Transition.'
85. Seema Alavi, 'Unani Medicine in the Nineteenth-Century Public Sphere: Urdu texts and the *Oudh Akhbar'*, *Indian Economic and Social History Review*, 42 (2005), 102–29, 4.
86. The role of vaccination within the medical apparatuses of colonial India is thoroughly explored in M. Harrison, *Public Health in British India: Anglo-Indian Preventive Medicine* 1859–1914 (Cambridge: Cambridge University Press, 1994), especially 82–87. At the same time, the experience of colonial inoculation campaigns contrasted starkly with an indigenous culture of treatment and affect around disease and its cures, especially in the context of smallpox. For an exploration of indigenous treatments of small pox in the context of vaccination campaigns, please see R.W. Nicholas, 'The Goddess Sitala and Epidemic Smallpox in Bengal', *Journal of Asian Studies*, 41, no. 1 (1981): 21–44. At the same time, Nigel Chancellor's brilliant recount of the use of propaganda in small pox campaigns in Mysore illuminates the political stakes of vaccination in Mysore. See N. Chancellor, 'A Picture of Health: The Dilemma of Gender and Status in the Iconography of Empire, India C. 1805', *Modern Asian Studies*, 35, no. 4 (2001): 769–82.
87. Please see Mike Davis, *Late Victorian Holocausts: El Niño Famines and the Making of the Third World* (Verso, 2002).
88. This understanding of the biopolitical owes a debt to David Arnold's pioneering work, which has set the scene for thinking through the connection between health, politics and everyday life in South Asia and other sites of empire. Please see, once more, Arnold, *Colonizing the Body: State Medicine and Epidemic Disease in Nineteenth-Century India*.
89. This phenomenon is well documented in the works of most medical historians. Guy Attewell, however, makes this point quite compellingly in relation to a plague outbreak in Delhi. Moving beyond the model of vaids or hakims coerced (or compelled to action), he argues that Unani tabeeb [0]involvement in vaccination campaigns was the result of a moral urgency experienced by hakims who were alarmed by the devastations of the plague, and adapted their techniques and technologies to respond to it. See Attewell, *Refiguring Unani Tibb*, chapter 3.
90. See L.N. Trivedi, 'Visually Mapping the 'Nation': Swadeshi Politics in Nationalist India, 1920–1930', *The Journal of Asian Studies*, 62, no. 1 (February 2003): 11–41.
91. C.A. Bayly, 'The Origins of Swadeshi (home industry): Cloth and Indian society, 1700–1930', in *The Social Life of Things*, ed. by Arjun Appadurai, (Cambridge: Cambridge University Press, 1986).; Lucy Norris, 'Shedding Skins: the Materiality of Divestment in India', *Journal of Material Culture*, 9 (2004), 59–71, pp. 59–61.
92. The most notable book on this topic is Alter, *Gandhi's Body: Sex, Diet and the Politics of Nationalism*. Echoes of Alter's work are felt in recent work by Jean Langford, Larry Cohen and Thomas Blom Hansen. See Lawrence Cohen, *No Aging in India: Alzheimer's, the bad family and other modern things* (Berkeley: 1998); Thomas Blom Hansen, *Saffron Wave: Democracy and Hindu nationalism in modern India* (Princeton: 1999); Langford, *Fluent Bodies: Ayurvedic Remedies for Postcolonial Imbalance*.

93. This is a recurrent theme in a lot of Gandhi's writings. It is reflected quite beautifully in his encounter with institutionalized Christianity in South Africa, recounted in Mahatma Gandhi, 'An autobiography; or, The story of my experiments with truth', in *The selected works of Mahatma Gandhi*, ed. Shriman Narayan, trans. Madhav Desai, vol. 1 (Ahmedabad: Navajivan Press, 1927), 181–85.

94. Vinay Lal has elegantly explored this very controversial part of Gandhi's life with intellectual and political integrity. Please see V. Lal, 'Nakedness, Non-Violence and the Negation of Negation: Gandhi's Experiments in Brahmachary and Celibate Sexuality', *South Asia*, 22, no. 2 (1999): 63–94. The moral virtue here is rendered somewhat problematic: though his intentions were honourable, historians have broadened the moral landscape of these actions by focusing on the exploitative possibilities associated with these behaviours. Madhu Kishwar has constructed an unparalleled feminist reading of Gandhi's actions vis-à-vis the women around him. Please see Madhu Kishwar, *Gandhi and Women* (Delhi: Manushi Prakashan, 1986).

## 2   Situating Ayurveda in Modernity, 1900–1919

1. Sarah Hodges, *Contraception, Colonialism and Commerce: Birth Control in South India, 1920–1940* (Aldershot: Ashgate, 2008), chapter 1.

2. A. Appadurai, 'Numbers in the Colonial Imagination', in *Orientalism and the Postcolonial Predicament: Perspectives on South Asia*, ed. P. Van de Veer and C. Breckenridge (Philadelphia: University of Pennsylvania Press, 1993). Dirks' work on the rigidification of caste identities as a product of colonial governance is the most prominent example of this argument. Please see N.B. Dirks, *Castes of Mind: Colonialism and the Making of Modern India* (Princeton: Princeton University Press, 2001). This is also a process we can identify in broader cultural terms that move beyond the boundary of the individual. David Lelyveld's work on the division between Hindi and Urdu in the nineteenth century reveals the extent to which the introduction of rigid colonial categories of language fuelled the divide between vernacular speakers. See D. Lelyveld, 'Colonial Knowledge and the Fate of Hindustani', *Comparative Studies in Society and History*, 35, no. 4 (1993): 665–682.

3. The prevailing models for understanding the history of medicine vis-à-vis the cementation of the state adopts a Foucauldian model of governmentality, as discussed in the Introduction to this book.

4. I come back again to Shula Marks in this instance, as we begin to see the other side of colonial medicine in the pre-existing institutions that laboured under the state but were not a part of its infrastructure. See S. Marks, 'What Is Colonial About Colonial Medicine? And What Has Happened to Imperialism and Health?', *Social History of Medicine*, 10, no. 02 (1997): 205–219.

5. See, for example, M. Lal, 'The Politics of Gender and Medicine in Colonial India: The Countess of Dufferin Fund, 1885–1888', *Bulletin of the History of Medicine*, 68 (1994): 29–66; G. Forbes, 'Managing Midwifery in India', in *Contesting Colonial Hegemonies: State and Society in Africa and India*, ed.

D. Engels and S. Marks (London: British Academy Press, 1994); S. Lang, 'Drop the Demon Dai: Maternal Mortality and the State in Madras Presidency 1840–1875' (2003); Samita Sen, 'Motherhood and Mothercraft: Gender and Nationalism in Bengal', *Gender & History*, 5, no. 2 (1993): 231–243.

6. Chakraborty, *Garbharaksha-a* (Calcutta, 1898).
7. Irfan Habib and Dhruv Raina have written about the incorporation of Western science into the curriculum at Hindu College in Delhi and also at Aligarh. I. Habib and D. Raina, *Domesticating Modern Science: Essays on Social History of Science and Culture in Colonial India* (New Delhi: Tulika, 2004).
8. Razzack, *Hakim Ajmal Khan, the Versatile Genius* (New Delhi: Central Council for Research in Unani Medicine, Ministry of Health and Family Welfare, 1987), 22.
9. Manmathnatha Datta, *Ayurveda; or the Hindu System of Medical Science* (Calcutta: H.C. Dass, 1899).
10. NAI Home (Medical) December 1895, nos. 15–18, 795.
11. NAI Home (Medical) December 1895, nos. 15–18, 789–790.
12. NAI Home (Medical) December 1895, nos. 15–18, 790.
13. NAI Home (Medical) December 1895, nos. 15–18, 791.
14. A. Kumar, *Medicine and the Raj: British Medical Policy in India, 1835–1911* (New Delhi: Sage, 1998), 160.
15. *Report of the Central Indigenous Drugs Committee*, 3 vols (1899), 1–3, p. 22. (Appendix 4).
16. *Report of the Central Indigenous Drugs Committee*, p. 22. (Appendix 4).
17. *Report of the Central Indigenous Drugs Committee*, p. 22. (Appendix 4).
18. NAI Home (Medical) May 1896, nos. 54–57, 601. Emphasis author's own.
19. *Report of the Central Indigenous Drugs Committee, vol. 1* (Government of India, Home Department, 1899), Appendices, 22 (appendix IV), p.37 (appendix & to Paras 21&39).
20. *Report of the Central Indigenous Drugs Committee, vol. 1*, Appendices, 22 (appendix IV). p. 44 of appendices (appendix X).
21. *Report of the Central Indigenous Drugs Committee, vol. 1*, Appendices, 22 (appendix IV). p. 44 of appendices (appendix X).
22. *Report of the Central Indigenous Drugs Committee, vol. 1*, Appendices, 22 (appendix IV). p. 44 of appendices (appendix X).
23. A brochure from 1942 shows a membership roster stating the different cities and alliances of vaids, stretching from Calcutta to Cochin to Delhi to Lahore.
24. Carey Watt's comprehensive study of the cultural of *seva* (service) in early twentieth century emphasizes the importance of these networks in the everyday life of community in North India. Please see C.A. Watt, *Serving the Nation: Cultures of Service, Association, and Citizenship in Colonial India* (Delhi: Oxford University Press, 2005).
25. Sanjay Joshi pushes further into this history in the context of social change in early twentieth-century Lucknow and argues that this process was crucial to the solidification of a middle class in India. Please see S. Joshi, *Fractured Modernity: Making of a Middle Class in Colonial North India* (Delhi: Oxford University Press, 2002).
26. UPSA Medical 3742/42, pamphlet from AIAC meeting, p. 1.
27. UPSA Medical 3742/42, pamphlet from AIAC meeting, p. 1.

28. Gananath Sen, *Pratyaksha Shariram* (Dehradun, 1929), sec. introduction.
29. Ibid., ii.
30. Ibid.
31. Ibid., iii.
32. Razzack, *Hakim Ajmal Khan, the Versatile Genius*, 10.
33. C.F. Andrews, *Hakim Ajmal Khan: A Sketch of His Life and Career* (Madras: G.A. Natesan and Co., 1922), 10.
34. File NAI Home (Medical), no. 130 (Nov 1914) deals with the request. Additional information about the allotment of land can be found in NAI Home (Medical), no. 129 (Jan 1914) and NAI Home (Land Revenue), no. 7–12 (Oct 1913).
35. NAI Home (Medical), no. 130 (Nov 1914), p. 1.
36. NAI Home (Medical), no. 130 (Nov 1914), p. 7–9.
37. NAI Home (Medical), no. 130 (Nov 1914), p. 2.
38. NAI Home (Medical), no. 130 (Nov 1914), p. 7. This claim makes vague reference to the conception of excess and waste associated with Indian tradition and ritual, thus further undermining these endeavours as scientific.
39. No.129, Imperial Delhi Committee to Government of India 24 January 2014, p. 1 in NAI Home (Medical), no. 130 (Nov 1914).
40. NAI Home (medical), no. 96–99a (April 1916).
41. NAI Home (Medical), no.8 (Aug 1912), 2.
42. NAI Home (Medical), no.8 (Aug 1912), Government of India to IMS 20 July 2012, 3.
43. NAI Home (Medical), no.8 (Aug 1912), Government of India to IMS 20 July 2012, 2.
44. NAI Home (Medical), no.8 (Aug 1912), Government of India to IMS 20 July 2012, 6.
45. The date for the laying of the stone was not confirmed in this correspondence, which was almost definitely a result of inadequate funding and planning permission. The stone was eventually laid by Viceroy Hardinge in 1916, four years later.
46. NAI Home (Medical) no, 174 (Dec 1915), p. 23.
47. NAI Home (Medical) no, 174 (Dec 1915), p. 23.
48. NAI Home (Medical) no. 174 (Dec 1915), p. 27.
49. NAI Home (Medical) no. 174 (Dec 1915), p. 32.
50. NAI Home (Medical) no. 22 (June 1917), p. 2.
51. NAI Home (Medical) no. 22 (June 1917), p. 3.
52. NAI Home (Medical) no.71–73 (Nov 1915), p. 1.
53. NAI Home (Medical) no.71–73 (Nov 1915), p. 27.
54. NAI Home (Medical) no.71–73 (Nov 1915), p. 5.
55. NAI Home (Medical) no.71–73 (Nov 1915), p. 12.
56. NAI Home (Medical) no. 38–50 (July 1916), Lukis to Craddock 8 January 2016.
57. NAI Home (Medical) no. 38–50 (July 1916),Legislative to Lukis 12 January 2016.
58. NAI Home (Medical) no. 38–50 (July 1916), p. 5–6.
59. NAI Home (Medical) no. 38–50 (July 1916), p. 5–6.
60. NAI Home (Medical) no. 26–50 (June 1919), p. 411.
61. NAI Home (Medical) no. 26–50 (June 1919), p. 411.

62. A final statement summarizing the position of the provinces can be found in NAI Home (Medical) (June 1919), no. 26–50A, 411.
63. NAI Home (Medical) 26–50A (June 1919), pp. 534–536.
64. NAI Education, Health and Lands (Health) April 73–75b, 1921.pp. 1–2.

## 3  Embodying Consumption: Representing Indigeneity in Popular Culture, 1910–1940

1. Please see V. Dalmia, *The Nationalization of Hindu Traditions: Bharatendu Harishchandra and Nineteenth Century Benaras* (Delhi: Oxford University Press, 1997); S. Joshi, *Fractured Modernity: Making of a Middle Class in Colonial North India* (Delhi: Oxford University Press, 2002); S. Kaviraj, *The Unhappy Consciousness: Bankimchandra Chattopadhyay and the Formation of Nationalist Discourse in India* (Delhi: Oxford University Press, 1995); F. Orsini, *The Hindi Public Sphere, 1920–1940: Language and Literature in the Age of Nationalism* (Delhi: Oxford University Press, 2002).
2. Orsini, *The Hindi Public Sphere, 1920–1940: Language and Literature in the Age of Nationalism*, 3; D. Lelyveld, 'Colonial Knowledge and the Fate of Hindustani', *Comparative Studies in Society and History*, 35, no. 4 (1993): 665–682.
3. Dalmia goes to great lengths to untangle the genealogies of different missionary works. See Dalmia, *The Nationalization of Hindu Traditions: Bharatendu Harishchandra and Nineteenth Century Benaras*, Chapter 1.
4. C.R. King, 'One Language, Two Scripts: The Hindi Movement in Nineteenth Century India' (1994).
5. Orsini, *The Hindi Public Sphere, 1920–1940: Language and Literature in the Age of Nationalism*, 69.
6. Lelyveld, 'Colonial Knowledge and the Fate of Hindustani'.
7. P.B. Mukharji, 'Enframing Bangali Ayurbed: Going Beyond "Frontier" Frames', *Journal of the Asiatic Society of Bangladesh (Humanities)*, 49, no. 1 (June 2004): 15. Translation from the Bengali is Mukharji's own. The reference is from Chondro, Obinash Kobirotno Kobiraj, *Chorok Shomhita*, Calcutta, 1941 (Vikram Samvat, 1884 A.D.).
8. Poonam Bala has written extensively on the first category of translation, which concerned itself with the canonical texts of Ayurveda. The use of the term 'eclectic' is Mukharji's, but suits my purposes well. The eclecticism he identifies refers to both the form and content of these publications. See P. Bala, *Imperialism and Medicine in Bengal: A Socio-historical Perspective* (New Delhi: Sage, 1991).
9. Orsini, *The Hindi Public Sphere, 1920–1940: Language and Literature in the Age of Nationalism*, 4.
10. Joshi, *Fractured Modernity: Making of a Middle Class in Colonial North India*, chapter Introduction.
11. For an in-depth take on the boundaries of belonging, especially around the categories of gender, sexuality and religion, see C. Gupta, *Sexuality, Obscenity, Community: Women, Muslims, and the Hindu Public in Colonial India* (Delhi: Permanent Black, 2002).

12. Michael Dodson has written about the appropriation of science as a key means of reconfiguring Sanskrit educational models in the mid-nineteenth century. See Michael S. Dodson, *Orientalism, Empire, and National Culture: India, 1770–1880* (Basingstoke: Palgrave Macmillan, 2007).The identification and appropriation of a 'Golden Age' of Hinduism was a common trope amongst various waves of reform. See, for example, U. Chakravarti, 'Whatever Happened to the Vedic Dasi? Orientalism, Nationalism and a Script for the Past', in *Recasting Women*, ed. K. Sangari and S. Vaid (New Delhi: Kali for Women, 1989); L. Mani, 'Contentious Traditions: The Debate on Sati in Colonial India', in *Recasting Women: Essays in Colonial History*, ed. K. Sangari and S. Vaid (Kali for Women, 1989), 88–126.

13. See Rohan Deb Roy, 'Debility, Diet, Desire: Food in Nineteenth and Early Twentieth Century Bengali Manuals', in *The Writer's Feast: Food and the Politics of Representation*, ed. Surpriya Chaudhuri and Rimi B Chatterjee (Hyderabad: Orient Blackswan, 2011), 179–205.

14. See, for example, Haridas Vaid, *Chikitsa Chandrodaya* (Allahabad, 1920).

15. Michael Dodson has explored this phenomenon as it pertained to scientific writings encountered, translated and disseminated by the pandits of Banaras. See Dodson, *Orientalism, Empire, and National Culture: India, 1770–1880*.

16. For an in-depth look at the evolution of Hindi into a language of nationalism, please see A. Rai, *Hindi Nationalism* (Delhi: Three Essays Collective, 2001).

17. The table of contents listed the two terms for fever in J. Sharma, *Arogya Darpan* (Prayag, 1915). Also see R. Sinha and D. Sinha, *Ayurvediya-Kosha/An Encyclopaedical Ayurvedic Dictionary* (Baralokpur-Ithava (United Provinces): Harihar Press, 1932), 3.

18. S. Sharma, *Carakasamhita* (Bombay: Adhyaksh Srivenkatekshar Press, 1931).

19. Ibid.

20. Francesca Orsini and Markus Daechsel both make compelling arguments about the consumption of literature amongst members of the Hindu and Muslim middle classes. Please see Orsini, *The Hindi Public Sphere, 1920–1940: Language and Literature in the Age of Nationalism*; M. Daechsel, *The Politics of Self-Expression: The Urdu Middle-class Milieu in Mid-twentieth Century India and Pakistan* (Abingdon: Routledge, 2006).

21. This is a point which other historians of the literary sphere in this period fail to make.

22. P.S. Shastri, *Ayurveda-mahattva* (Lucknow, 1925).

23. Vaid, *Chikitsa Chandrodaya*.

24. On the importance of visual markers of racial and ethnic identification as reflected through clothing, see E. Tarlo, *Clothing Matters: Dress & Its Symbolism in Modern India* (Chicago: University of Chicago Press, 1996); E.M. Collingham, *Imperial Bodies: The Physical Experience of the Raj, 1800–1947* (Cambridge: Polity Press, 2001).

25. R. Sinha and D. Sinha, *Ayurvediya-kosha: An Encyclopaedical Ayurvedic Dictionary (with Full Details of Ayurveda, Unani and Allopathic terms)* (Baralokpur, 1934), title page.

26. Ibid.

27. Ibid.

28. G. Prakash, *Another Reason: Science and the Imagination of Modern India* (Princeton: Princeton University Press, 1999), 9.

29. Sharma, *Carakasamhita*, 13.

30. Shastri, *Ayurveda-mahattva*, title page.

31. Editorial (no author listed), *Chand*, July 1930, 32.

32. See M. Hancock, 'Home Science and the Nationalization of Domesticity in Colonial India', *Modern Asian Studies*, 35, no. 4 (2001): 871–903.

33. See J. Bagchi, 'Representing Nationalism: Ideology of Motherhood in Colonial Bengal', *Economic and Political Weekly*, 20 October, 1990; I. Chowdhury-Sengupta, 'Mother India and Mother Victoria: Motherhood and Nationalism in Nineteenth-Century Bengal', *South Asia Research* 1, no. May (1992): chapter 3.

34. Prakash, *Another Reason: Science and the Imagination of Modern India*, 147.

35. P. Chatterjee, 'The Nationalist Resolution of the Women's Question', in *Recasting Women: Essays in Colonial History*, ed. K. Sangari and S. Vaid (Delhi: Kali for Women Press, 1989), 239.

36. L. Carroll, 'Law, Custom, and Statutory Social Reform: The Hindu Widow's Remarriage Act of 1856', in *Women in Colonial India: Essays on Survival, Work, and State*, ed. J. Krishnamurthi (Oxford: Oxford University Press, 1989), 1–26; P.A. McGinn, 'The Age of Consent Act (1891) Reconsidered: Women's Perspectives and Participation in the Child Marriage Controversy in India', *South Asia Research*, 12, no. 2 (1992): 100–118; M. Sinha, 'The Lineage of the "Indian" Modern: Rhetoric, Agency and the Sarda Act in Late Colonial India', in *Gender, Sexuality and Colonial Masculinities*, ed. A. Burton (New York, 1999), 207–221; P. Anagol, 'The Emergence of the Female Criminal in India: Infanticide and Survival Under the Raj', *Indian Economic and Social History Review*, 53, no. 1 (Spring 2002); M. Kasturi, 'Law and Crime in India: British Policy and the Female Infanticide Act of 1870', *Indian Journal of Gender Studies*, 1, no. 2 (1994): 169–194.

37. For an exploration of the role of medicine in mediating discussions of sexuality, see Gupta, *Sexuality, Obscenity, Community: Women, Muslims, and the Hindu Public in Colonial India*, chapter 4.

38. Hodges' notion of the counter-discourse of the mother-in-law shows the vilification of this figure as a backward entity, whose authority needed to be replaced by that of the younger, educated and modern women of the household. See S. Hodges, 'Governmentality, Population and Reproductive Family in Modern India', *Economic and Political Weekly* (2004): 1157–1163.

39. The term is Yashoda Devi's. However, even authors like Premchand, who were so critical of the elitism in this discourse, employed it as well (albeit mockingly at times). See Premchand, 'Mahila-sabhaon Mein Santan-nigrah Ka Prastav', in *Vividh Prasang*, ed. Premchand (Allahabad, 1932), 252.

40. M.K. Gandhi, *Arogya Sadhana* (Calcutta, 1922), 22.

41. On the role of midwives in the infanticide process, see Kasturi, 'Law and Crime in India: British Policy and the Female Infanticide Act of 1870'. C.A. Bayly also remarks on the importance of the midwife as an informal informant with regard to the spreading of information in colonial India. C.A. Bayly, *Empire and Information: Intelligence Gathering and Social Communication in India, 1780–1870* (Cambridge: CUP, 1996), 19. Furthermore, the

testimony of the midwife was solicited, especially in Bengal, in cases ranging from rape allegations to infanticide, which once again highlights the importance of the *dai* as medical informer; see Samita Sen, 'Motherhood and Mothercraft: Gender and Nationalism in Bengal', *Gender & History*, 5, no. 2 (1993): 231–243.

42. G. Forbes, 'Managing Midwifery in India', in *Contesting Colonial Hegemonies: State and Society in Africa and India*, ed. D. Engels and S. Marks (London: British Academy Press, 1994). on the role of midwives in the infanticide process

43. See Chakraborty, *Garbharaksha-a* (Calcutta, 1898).

44. 'Garbh-Pira ka upay' in Y. Devi, *Dampati Arogyata Jivanshastra* (Allahabad, 1931).

45. Gurunncharay, *Hindu Mata Athart Hindu Shishu Svastya Raksha (Hindu Streeyon Ke Liye Shishu Svastya Aur Shishu Sambandhi Pustak)* (Lahore, 1926); S. Sharma, *Shishu Palan* (Banaras, 1933).

46. R. Sharma, *Dugdh Chikitsa* (Banaras, 1918), 1.

47. Ibid., 2.

48. 'Santan-Vidhi Nigra' Sushiladevi, *Dampatya Jeevan* (1930), 139–140.

49. Charu Gupta has written a definitive article that provides an overview of Yashoda Devi's personal life and career that reflects upon her relationship to Ayurveda and to the transformation of middle-class women's lives in the early twentieth century. Please see Charu Gupta, 'Procreation and Pleasure Writings of a Woman Ayurvedic Practitioner in Colonial North India,' *Studies in History*, 21, no. 1 (February 2005): 17–44.

50. Yashoda Devi was author of over 40 guides, mostly famously including Yashoda Devi, *Arogya-Vidhana-grihini-kartavya Sastra* (Allahabad: 1924); Yashoda Devi, *Dampati Arogyata Jivanshastra Artha Ratishastra* (Allahabad: 1924–1925);Yashoda Devi, *Kanya-Karttavya* (Allahabad: 1925);Yashoda Devi, *Pati Prema Patrika* (Allahabad: 1925).

51. Maneesha Lal explores the particular obstacles facing women medical doctors in India, as well as the larger politics of gender and healthcare. See M. Lal, 'The Ignorance of Women Is the House of Illness': Gender, Nationalism, and Health Reform in Colonial North India', in *Medicine and Colonialism: The Politics of Identity*, ed. M.P. Sutphen (London, 2003).

52. For a study of the origins of women-centred healthcare in India, including the emergence of women in the medical profession in the subcontinent, see M. Lal, 'The Politics of Gender and Medicine in Colonial India: The Countess of Dufferin Fund, 1885–1888', *Bulletin of the History of Medicine*, 68 (1994): 29 66; A. Burton, 'Contesting the Zenana: The Mission to Make "Lady Doctors for India", 1874–1885', *The Journal of British Studies*, 35, no. 3 (1996): 368–397.

53. Yashoda Devi, *Dampati Arogyata Jivanshastra* (Allahabad, 1931), 77.

54. This is a line of argument that historians interested in eugenics in colonial settings have recently begun to identify as a rather consistent mode of reasoning. See S. Hodges, *Reproductive Health in India: History, Politics, Controversies*, vol. 13 (Hyderabad: Orient Longman, 2006), http://wrap.warwick.ac.uk/38898/.

55. See Joshi, *Fractured Modernity: Making of a Middle Class in Colonial North India*, introduction.

56. See Gupta, *Sexuality, Obscenity, Community: Women, Muslims, and the Hindu Public in Colonial India*, 72–83. Gupta discusses these advertisements in the context of the banning of obscenity and the related regulation of sexuality and thinks through advertising of cures for sexual ailments as another mode of this process.

57. For the history of the lock hospital and other means through which the colonial state attempted to control venereal disease, please see D.M. Peers, 'Privates Off Parade: Regiments and Sexuality in the Nineteenth Century Indian Empire', *International History, Review*, 20, no. 4 (1998): 823–854; Sarah Hodges, 'Looting the Lock Hospital in Colonial Madras During the Famine Years of the 1870s', *Social History of Medicine*, 18, no. 3 (2005): 379–398; P. Levine, *Prostitution, Race and Politics: Policing Venereal Disease in the British Empire* (London: Routledge, 2003). A remedy for gonorrhoea can be found in Gupta, *Pak-Prakash athava Mithai* (Pratapgarh, 1930), a popular cookbook that circulated from the 1930s.

58. Ad for Gonokiller, *Madhuri*, November 1940, 213.

59. Joseph Alter and Carey Watt both discuss the importance of sports and physicality to the performance of national identity in the late nationalist period. See J. Alter, *The Wrestler's Body: Identity and Ideology in North India* (Berkeley: University of California Press, 1992); C.A. Watt, *Serving the Nation: Cultures of Service, Association, and Citizenship in Colonial India* (Delhi: Oxford University Press, 2005).

60. Ad found in *Stree Darpan*, January 1936, 17.

61. Ad for Vita-milk found in *Saraswati*, March 1929, 13.

# 4   Ayurveda's Dyarchic Moment, 1920–1935

1. Shula Marks' perennial question about the categorization of 'colonial' medical techniques is particularly relevant in this context.

2. G. Pandey, *The Ascendancy of the Congress in Uttar Pradesh: Class, Community and Nation in Northern India, 1920–1940* (London: Anthem, 2002), chapter 1.

3. See Lionel Curtis, *Papers Relating to the Application of the Principle of Dyarchy to the Government of India: To Which Are Appended the Report of the Joint Select Committee and the Government of India Act, 1919* (Oxford: The Clarendon Press, 1920), xxi–xxii. This decision was most emblematically represented by the Lucknow Pact of 1916.

4. Ibid., xxiii.

5. The figure is found in David Arnold, 'The Armed Police and Colonial Rule in South India, 1914–1947', *Modern Asian Studies*, 11, no. 1 (1977): 103.

6. The work most commonly associated with this position are A. Seal, *The Emergence of Indian Nationalism: Competition and Collaboration in the Later Nineteenth Century*, vol. 1 (Cambridge: Cambridge University Press, 1971); J. Gallagher, G. Johnson, and A. Seal, *Locality, Province and Nation: Essays on Indian Politics, 1870–1940* (London: CUP, 1973). Michelguglielmo Torri evaluates the legacy of the shifting discourse on class in M. Torri, ' "Westernized Middle-Class", Intellectuals and Society in Late Colonial India', in *The Congress and Indian Nationalism: Historical Perspectives*, ed. J.L. Hill (London: Curzon, 1991).

7. S. Joshi, *Fractured Modernity: Making of a Middle Class in Colonial North India* (Delhi: Oxford University Press, 2002).
8. Sarah Hodges, *Contraception, Colonialism and Commerce: Birth Control in South India, 1920–1940* (Aldershot: Ashgate, 2008), chapter 1.
9. Legg's recent work on scalar geographies in relation to dyarchy span several articles. Please see Stephen Legg, 'Of Scales, Networks and Assemblages: The League of Nations Apparatus and the Scalar Sovereignty of the Government of India', *Transactions of the Institute of British Geographers*, 34, no. 2 (2009): 234–253; Stephen Legg, 'Stimulation, Segregation and Scandal: Geographies of Prostitution Regulation in British India, Between Registration (1888) and Suppression (1923)', *Modern Asian Studies*, 46, no. 6 (2012): 1459–1505; Stephen Legg, 'Transnationalism and the Scalar Politics of Imperialism', *New Global Studies*, 4, no. 1 (2010): 1–7; Stephen Legg, 'Planning Social Hygiene: From Contamination to Contagion in Interwar India', in *Imperial Contagions: Medicine, Hygiene, and Cultures of Planning in Asia*, ed. Robert Peckham (Hong Kong: Hong Kong University Press, 2013), 105–122.
10. This understanding of the function of scalar politics as a model is detailed in Legg, 'Of Scales, Networks and Assemblages', 235.
11. M. Harrison, *Public Health in British India: Anglo-Indian Preventive Medicine 1859–1914* (Cambridge: Cambridge University Press, 1994); David Arnold, *Colonizing the Body: State Medicine and Epidemic Disease in Nineteenth-Century India* (Berkeley: University of California Press, 1993); A. Kumar, *Medicine and the Raj: British Medical Policy in India, 1835–1911* (New Delhi: Sage, 1998).
12. This, is of course, David Arnold's term, though non-Foucauldians like Mark Harrison and Deepak Kumar concur with this assessment, and it has become the norm in discussions of science and colonialism.
13. Arnold, *Colonizing the Body: State Medicine and Epidemic Disease in Nineteenth-Century India*, 10.
14. See Harrison, *Public Health in British India: Anglo-Indian Preventive Medicine 1859–1914*, chapter 8. While Harrison details the myriad ways in which independent public health schemes were ordered by the district commissioners to fit the needs and capabilities of the provinces, the logic behind the sanitation reforms that local municipal leaders were meant to implement makes evident the hand of the central government in creating uniform policies for the subcontinent.
15. Uttar Pradesh State Archives (UPSA), Medical file 165 of 1923, p. 1.
16. Uttar Pradesh State Archives (UPSA), Medical file 165 of 1923, p. 2.
17. Uttar Pradesh State Archives (UPSA), Medical file 165 of 1923, p. 3.
18. Uttar Pradesh State Archives (UPSA), Medical file 165 of 1923, p. 7.
19. Uttar Pradesh State Archives (UPSA), Medical file 165 of 1923, p. 11.
20. Uttar Pradesh State Archives (UPSA), Medical file 165 of 1923, p. 12.
21. This issue was reportedly the first one to come before this body, which was formed after the transfer of medical decision-making in 1921–1922. See Uttar Pradesh State Archives (UPSA), Medical file 165 of 1923, p. 14.
22. Uttar Pradesh State Archives (UPSA), Medical file 165 of 1923, p. 16.
23. Uttar Pradesh State Archives (UPSA), Medical file 165 of 1923, p. 23.
24. Uttar Pradesh State Archives (UPSA), Medical file 165 of 1923, p. 26.
25. Uttar Pradesh State Archives (UPSA), Medical file 165 of 1923, p. 27.

26. Uttar Pradesh State Archives (UPSA), Medical file 165 of 1923, p. 31. This refers to the Vaid Samelan in Cawnpore and the Tikmil-ul Tibb in Lucknow.
27. Interestingly, here, he compares practitioners to engineers, stating that the latter had experience on the ground and with building materials, and therefore had developed an expertise that need not be taught or regulated. Practitioners, he argues quite conversely, have much less experience with substances despite long periods of practice or use.
28. UPSA, Medical file 165 of 1923, p. 33.
29. UPSA, Medical file 165 of 1923, Resolution no. 137 of 1927, p. 43 of file.
30. UPSA, Medical file 165 of 1923, p. 44.
31. UPSA, Medical file 165 of 1923, p. 46.
32. UPSA, Medical file 165 of 1923, p. 50.
33. UPSA Medical file 495, 1925, p. 1.
34. Muir to Hon'ble Minister of Education, UPSA Medical file 495, 1925.
35. UPSA Medical 131/1933, box 59, p. 1.
36. See, for example, Paul Breton or Thomas Wise's nineteenth-century tracts on indigenous medicine. Both laud the ancient chemistry, physics and anatomy mentioned in the texts of Ayurveda they encountered. See Thomas Alexander Wise, *Thomas Alexander, Commentary on the Hindu System of Medicine* (Calcutta: Thacker, 1845); Peter Breton, *A Vocabulary of the Names of the Various Parts of the Human Body and of Medical and Technical Terms in English, Arabic, Persian, Hindee and Sanscrit for the Use of the Members of the Medical Department in India* (Calcutta: Government Lithographic Press, 1825).
37. UPSA Medical 131/1933, box 59, p. 3–5.
38. UPSA Medical 131/1933, box 59, p. 4.
39. UPSA Medical 131/1933, box 59, p. 4.
40. UPSA Medical 131/1933, box 59, p. 7.
41. UPSA Medical 131/1933, box 59, p. 10.
42. S.P. Shah to N.R. Sarkar, UPSA Medical 131/1933, box 59, p. 19.

# 5   Planning Through Development: Institutions, Population and the Limits of Belonging

1. Benjamin Zachariah, *Playing the Nation Game: The Ambiguities of Nationalism in India* (Calcutta: Yoda Press, 2012).
2. Two recent works on health during the period before and after decolonization in the subcontinent both make this point about the Central Government's ideological shift in this period. Please see S.S. Amrith, *Decolonizing International Health: India an Southeast Asia, 1930–1965* (Basingstoke: Palgrave Macmillan, 2006); S. Bhattacharya, *Expunging Variola: The Control and Eradication of Smallpox in India, 1947–1977* (London: Sangam, 2006).
3. G. Pandey, *The Ascendancy of the Congress in Uttar Pradesh: Class, Community and Nation in Northern India, 1920–1940* (London: Anthem, 2002), 28–29.
4. Ibid., 35.
5. D.A. Washbrook, 'The Rhetoric of Democracy and Development in Late Colonial India', in *Nationalism, Democracy and Development*, ed. S. Bose and A. Jalal (Delhi: OUP, 1997), 42.

6. P.R. Brass, *Language, Religion and Politics in North India* (Cambridge: CUP, 1974); P.R. Brass, 'A Reply to Francis Robinson', *Journal of Commonwealth and Comparative Politics*, XV (1977): 231–233; F. Robinson, 'Nation Formation: The Brass Thesis and Muslim Separatism', *Journal of Commonwealth and Comparative Politics* XV (1977): 215–230.

7. See, for instance, B.R. Tomlinson, *The Political Economy of the Raj, 1914–1947: The Economics of Decolonization in India* (London: Macmillan, 1979).

8. On the intricacies of central rule as it played out in the provincial police and military forces, please see T. C. Sherman, *State Violence and Punishment in India* (London: Routledge, 2009).

9. UPSA Medical 165/1923, p. 5.

10. UPSA Medical 165/1923, p. 29.

11. UPSA Medical 165/1923, p. 2.

12. UPSA Medical 165/1923, p. 3.

13. UPSA Medical 236/1937,G.B. Pant to V.L. Pandit, 1 June 1938, p.11 in proceedings.

14. UPSA Medical 236/1937, p 1–2.

15. UPSA Medical 236/1938, p. 32.

16. UPSA Medical 236/1938, p. 32.

17. UPSA Medical 236/1938, p. 32.

18. See, for example, Pandey, *The Ascendancy of the Congress in Uttar Pradesh: Class, Community and Nation in Northern India, 1920–1940*; W. Gould, *Hindu Nationalism and the Language of Politics in Late Colonial India* (Cambridge: Cambridge University Press, 2004). Gould in particular creates a significant model for evaluating the social and cultural influence of the rhetoric employed by the Congress in this period.

19. UPSA Medical 236/1938, p. 1.

20. UPSA Medical 236/1938, p. 113.

21. UPSA Medical 236/1938, p. 136.

22. Arjun Appadurai and others have noted the importance of data collection through the census and by other means as integral to the imperial project through the use of surveillance techniques and evaluative technologies of 'modern' – read, Western, – governance. See A. Appadurai, 'Numbers in the Colonial Imagination', in *Orientalism and the Postcolonial Predicament: Perspectives on South Asia*, ed. P. Van de Veer and C. Breckenridge (Philadelphia: University of Pennsylvania Press, 1993).

23. UPSA Medical 26/1938, p. 3.

24. UPSA Medical 26/1938, p. 8. The Ayurvedic Pharmacy, Gurukula University in Haridwar and the Ayurvedic Pharmacy, Bundhelkand Ayurvedic College, Jhansi were added in late 1939 as well p. 10.

25. UPSA Medical 26/1938, pp. 4–6.

26. UPSA Medical 26/1938, p. 16. Interestingly, the Tibbia College in Delhi was allowed to remain on the list without any explanation. Included in this set of files was a glowing letter of recommendation for Hakim Jamil Khan of Delhi, a descendant of Ajmal Khan and the director of the Tibbia College in the late 1930s (1–2). The Board of Indian Medicine also felt that the college was respected by both Hindus and Muslims, making it an appropriate dispensary for both Ayurvedic and Unani medicine p. 16.

27. UPSA Medical 26/1938, pp. 33–35.

28. UPSA Medical 26/1938, p. 43.
29. UPSA Medical 26/1938, p. 44.
30. Srivastava to Elliott, 3-5-40, UPSA Medical 26/1938.
31. Though he towed the government line by agreeing that indigenous practitioners should be appointed after the war (Srivastava to Elliott, 3-5-40, UPSA Medical 26/1938).
32. UPSA Medical 155/1938, pp. 1–2.
33. UPSA Medical 155/1938, p. 9.
34. UPSA Medical 155/1938, p. 18.
35. UPSA Medical 155/1938, p. 26.
36. UPSA Medical 155/1938, p. 62.
37. UPSA Medical 155/1938, p. 41.
38. UPSA Medical 155/1938, pp. 66–67.
39. UPSA Medical 155/1938, pp. 68–69.
40. Kavita Sivaramakrishnan, *Old Potions, New Bottles: Recasting Indigenous Medicine in Colonial Punjab* (Hyderabad: Orient Longman, 2006).
41. Seema Alavi, 'Unani Medicine in the Nineteenth-century Public Sphere: Urdu Texts and the Oudh Akhbar,' *Indian Economic and Social History Review*, 42, no. 1 (2005): 102–129.
42. See Guy Attewell, *Refiguring Unani Tibb* (Hyderabad: Orient Longman, 2007), chapter 4.
43. Jean Langford, *Fluent Bodies: Ayurvedic Remedies for Postcolonial Imbalance* (Durham: Duke University Press, 2002), chapters 2 & 3.
44. See J. Sharma, *Arogya Darpan* (Prayag, 1915). In it, Sharma tells the story of his family's ongoing practice in the UP throughout the nineteenth century.
45. UPSA Medical 116/1941, pp. 2–3.
46. UPSA Medical 116/1941, p. 3.
47. See Leah Renold's work on the history of BHU, in which she positions the institution as the cultural and political centre of UP politics. L. Renold, *A Hindu Education: Early Years of Banaras Hindu University* (New Delhi: OUP, 2005).
48. UPSA Medical 250/1926, p. 2.
49. UPSA Medical 1088/48, p. 1.
50. UPSA Medical 1088/48, p. 1.
51. UPSA Medical 1088/48, p. 1.
52. UPSA Medical 1088/48, p. 1.
53. UPSA Medical 1088/48, p. 1.
54. UPSA Medical 1088/48, p. 2.
55. UPSA Medical 1088/48, pp. 3–4.
56. This became more and more evident in the 1940s, when the BIM takes on more responsibility.
57. UPSA Medical 250/1926, Justice Gokaran Nath Misra to Sir Ivo Elliott, 20 January 1928.
58. UPSA Medical 250/1926, pp. 18–20 & UPSA Medical 250/1926, p. 31.
59. UPSA Medical 250/1926, pp. 53–55, Justice Gokaran Nath Misra to Sir Ivo Elliott, 16 March 1928, p.1.
60. UPSA Medical 1088/48, p. 3.
61. See, for example, the records pertaining to the audit of the Takmil-ul-Tibb which counted amongst its governors several prominent Hindus and vaids (UPSA Medical 14/1942 box 179).

62. UPSA Medical 428/1940 box 99, p. 11.
63. UPSA Medical 428/1940 box 99, p. 13.
64. UPSA Medical 428/1940 box 99, p. 14.
65. UPSA Medical 428/1940 box 99, p. 15.
66. UPSA Medical 428/1940 box 99, p. 1.
67. UPSA Medical 428/1940 box 99, p. 2.
68. UPSA Medical 428/1940 box 99, p. 11.
69. UPSA Medical 428/1940 box 99, p. 10.
70. UPSA Medical 428/1940 box 99, p.12.

## 6 Reframing Indigeneity: Ayurveda, Independence and the Health of the Future

1. I use the term independence advisedly here, differentiating between the postcolonial and the independence in the vain of criticism raised by Ella Shohat; see E. Shohat, 'Notes on the Post-Colonial', *Social Text* no. 31/32 (1992): 99–113. Shohat thinks through the ahistorical and totalizing effects of any invocation of the postcolonial, therein limiting the limits of the political agency by shutting off a further quest to delve into the meaning of specific conditions of experience. I use independence instead to reflect the optimism of the contemporary actors, while also better situating the political affectations built around the idea of an independent India at the forefront of discussions of the playing out of politics.
2. Please see S. Bose, *His Majesty's Opponent: Subhas Chandra Bose and India's Struggle Against Empire* (Cambridge: Harvard University Press, 2011).
3. Yasmin Khan's thorough history of partition provides an excellent and up-to-date summation of this tumultuous time in India's history. Please see Yasmin Khan, *The Great Partition: The Making of India and Pakistan* (New Haven: Yale University Press, 2007).
4. Shabnum Tejani's work on secularism in late empire explores the liberal roots of late congress ideology. Please see Shabnum Tejani, *Indian Secularism: A Social and Intellectual History, 1890–1950* (Bloomington: Indiana University Press, 2008).
5. Sunil Amrith makes evident the diversity of the committee, which included several conservative members, along with two openly communist ones, and makes the case that this diversity of worldviews likely led to the radical shift in planning proposed by the committee. Amrith also attributes some of the more careful findings of the report to discussions held with a group of touring international consultants who were in India on a trip sponsored by the Rockefeller Foundation. Please see S.S. Amrith, *Decolonizing International Health: India and Southeast Asia, 1930–1965* (Basingstoke: Palgrave Macmillan, 2006), 57–63.
6. J. Bhore, 'Report of the Health Survey and Development Committee', *GOI*, *New Delhi* (1946), vol. 1, 6–7.
7. Ibid., 3.
8. Ibid., 12.
9. Ibid., 12.
10. In the proposed plan for the deployment of health services, the areas addressed would be redrawn by the number of people living within it, taking

3,000,000 as limit for a division, with subdivisions based on a figure of 10,000–20,000 people, and officers would be deployed accordingly. Please see Ibid., vol. 2, 38.

11. Historians of gender and social life have been making this argument about the role of the state and other public institutions in the everyday spaces of empire for several decades, as we saw in Chapter 3.

12. Even as late as 1993, the Bhore Report remained the most thorough and progressive survey of India's health-care system. Please see N.H. Antia, 'An Alternative Strategy for Health Care? The Mandwa Project', *Economic and Political Weekly*, 20, no. 51/52, 21 December 1985: 2257–2260.

13. The historiography of social hygiene is found across the works of several historians and historical geographers, most of who work, in some measure, within the domain of the history of sexuality. Social hygiene was best articulated in relation to the threats posed to it by sexual deviance, in the form of sex work (and its labourers) and also in the realm of non-/normative conjugal relations, especially with regard to questions of birth control, population control and eugenics. These themes are best discussed in B. Ramusack, 'Embattled Advocates: The Debate Over Birth Control in India,1920–1940', *Journal of Women's History*, 1, no. 2 (1989): 34–64; Sarah Hodges, *Contraception, Colonialism and Commerce: Birth Control in South India, 1920–1940* (Aldershot: Ashgate, 2008); S. Hodges, 'Governmentality, Population and Reproductive Family in Modern India', *Economic and Political Weekly* (2004): 1157–1163; Stephen Legg, 'Planning Social Hygiene: From Contamination to Contagion in Interwar India', in *Imperial Contagions: Medicine, Hygiene, and Cultures of Planning in Asia*, ed. Robert Peckham (Hong Kong: Hong Kong University Press, 2013), 105–122; Sanjam Ahluwalia, *Reproductive Restraints: Birth Control in India, 1877–1947* (University of Illinois Pr, 2007); D. Arnold, 'Official Attitudes to Population, Birth Control and Reproductive Health in India, 1921–46', in *Reproductive Health in India: History, Politics, Controversies*, ed. S.E. Hodges (Delhi, 2006); Alison Bashford, *Imperial Hygiene: A Critical History of Colonialism, Nationalism and Public Health* (Basingstoke: Palgrave Macmillan, 2004).

14. Please see Legg's elaboration of this phenomena from the perspective of planning in Legg, 'Planning Social Hygiene: From Contamination to Contagion in Interwar India'.

15. Douglas Haynes and Shrikant Botre are currently engaged in work on Dr R.D. Karve, early twentieth-century sexologist engaged with both local Gujarati and international audiences. See D. Haynes and S. Botre, 'R.D. Karve and Male Middle-Class Sexuality in Western India: Correspondence Published in Samaj Swasthya, 1927–53', forthcoming.

16. *Report of the Committee on Indigenous Systems of Medicine*, p. 1. The committee also expanded this knowledge to include human dissection, the study of organ function, bacteriology, immunology and methods of preventative illness.

17. R.N. Chopra, *Report of the Committee on Indigenous Systems of Medicine*, Chopra Report, n.d., India Office Records.

18. Ibid., 3.

19. Ibid., 3–4.

20. Ibid., 4. This date is incorrect, mistaken for the introduction of Macaulay's Minute on Education of 1835, which put an end to the vernacular education in state-funded institutions, as discussed in Chapter 2.
21. Ibid., 6. The figure of 80% was itself taken from a speculative comment in the Bhore Report.
22. These figures and modes of conceptualizing the health of the nation were lifted directly from the Bhore Report, which is cited extensively within the Chopra Report. R.N. Chopra, *Report of the Committee on Indigenous Systems of Medicine* (Technical report. New Delhi: Ministry of Health, Government of India. The Chopra Report, 1948), 68.
23. Chopra, *Report of the Committee on Indigenous Systems of Medicine*, n.d., 71.
24. This citation originates with the Bhore, with emphasis added in the original. Ibid.
25. Ibid., 81. The attempts to identify a pro-philosophy cadre amongst the biomedical community, while almost comical in their execution, reveal the limitations of this kind of mission.
26. Ibid., 84.
27. Ibid., 121.
28. This argument was made most convincingly by Partha Chatterjee, as discussed earlier. See Partha Chatterjee, *The Nation and Its Fragments: Colonial and Postcolonial Histories* (Princeton: Princeton University Press, 1993), 13–34.
29. UPSA Medical B 38RC/49 box 32, p. 3.
30. UPSA Medical B 38RC/49 box 32, p. No. 7313/2062, letter from Dr. B.D. Wadhwa, Director of Health services (Ayurveda) Lucknow, to The Chairman, BIM, Lucknow, p. 1.
31. UPSA Medical Medical 19RC/1949, No. 7313/2062, letter from Dr. B.D. Wadhwa, Director of Health services (Ayurveda) Lucknow, to The Chairman, BIM, Lucknow, p. 2. The district officers were to be paid Rs 10/lecture and Rs 300/demonstration, an amount they'd accrue from University coffers which would accumulate on top of their regular salary.
32. UPSA Medical B 38RC/49 box 32, pp. 30–33.
33. UPSA Medical B 38RC/49 box 32, p. 22.
34. UPSA UPSA Medical B 38RC/49 box 32, p. 40.
35. UPSA Medical 22RC/49, 1949, p. 1.
36. UPSA Medical 22RC/49, 1949, p. 3.
37. UPSA Medical 22RC/49, p. 5.
38. UPSA Medical 22RC/49, p. 5.
39. UPSA Medical 22RC/49, p. 6.
40. UPSA Medical 25RC/49 box 29, p. 39.
41. UPSA Medical 25RC/49 box 29, p. 38.
42. UPSA Medical 25RC/49 box 29, p. 17.
43. UPSA Medical 25RC/49 box 29, p. 10.
44. UPSA Medical 25RC/49 box 29, p. 10.
45. UPSA Medical 25RC/49 box 29, p. 12.
46. UPSA Medical 25RC/49 box 29, p. 15.
47. UPSA Medical 2RC/49 box 205, p. 1.
48. Sarma, *Paka-Vijnana* (Benares: 1933).

49. See Arjun Appadurai, 'How to Make a National Cuisine: Cookbooks in Modern India', *Comparative Studies in Society and History,* 30 (1988), 3–24.
50. Gupta, *Pak-Prakash athava Mithai.*
51. Yashoda Devi, *Grihini Kartavya Sastra* (Allahabad: 1915–1916).
52. Ad for Dalda spice paste found in *Madhuri,* January 1943, p. 38.
53. E.M. Collingham, *Curry: a Tale of Cooks and Conquerors* (London: Vintage Books, 2005), 190–191. Collingham also notes that tea was the prime commodity imported from China between 1811–1819, but was an unreliable commodity, as tea was grown on private, family-owned plantations and so its availability was haphazard.
54. Ibid., 196.
55. Bharat Chai Tea Company ad, *Stri Darpan,* July 1937, p. 42.
56. Hakim Maulvia Muhammad Abudllah Saheb, *Bargad Ke Goon Tatha Uske Upyog* (Doheta Pustak Bazaar, 1944), 3.

# Bibliography

## Official Records

### Oriental and India Office Collections, London (IOR)
Board of Revenue Collection
Department Annual Reports
Medical
Education
Sanitary
Home Proceedings
Medical Department
Education, Health and Lands Department
Sanitary Department

### National Archives of India, New Delhi (NAI)
Government of India, Home Department files:
Medical Department
Education, Health and Lands Department
Sanitary Department

### United Provinces State Archives, Lucknow (UPSA)
Proceedings and Files of the Government of India:
Education Department
General Administration Department
Medical Department
Sanitary Department

## Periodicals and Newspapers

### Hindi
Stree Darpan
Chand
Madhuri
Saraswati
Roshni
Hans

### English
Asiatick Researches
Journal of the Royal Asiatic Society of Bengal

Transactions of the Medical and Physical Society of Bengal
The Lancet
The British Medical Journal

## Primary Printed Materials

### English

———— *Report of the Central Indigenous Drugs Committee*, 3 vols (1899).

———— *Report of the Health Survey and Development Committee*, 4 vols (1943).

———— *Report of the Committee on Indigenous Systems of Medicine*, 2 vols (1946).

Andrews, C. F. *Hakim Ajmal Khan: A sketch of his life and career* (Madras: 1922).

Babur, *The Babur-Nama in English: Memoirs of Bābur*, trans. by Annette Beveridge (London: 1969).

Bernier, Francois. *Travels in the Mogul Empire*, trans. by Irving Brock (London: 1826).

Breton, Peter. *A Vocabulary of the Names of the various parts of the Human Body and of Medical and Technical Terms in English, Arabic, Persian, Hindee and Sanskrit for the use of the Members of the Medical Department in India* (Calcutta: 1825).

Curtis, Lionel. *Papers Relating to the Application of the Principle of Dyarchy to the Government of India: To Which Are Appended the Report of the Joint Select Committee and the Government of India Act, 1919* (The Clarendon Press: 1920).

Datta, Manmathnatha. *Ayurveda; or the Hindu System of Medical Science* (Calcutta: 1899).

Fazl, Abul, *The Ain-iAkbari by Abul Fazl-i-Allami*, trans. by H.S. Jarrett (Calcutta: 1894).

Gandhi, M.K. "An autobiography; or, The story of my experiments with truth." in *The Selected Works of Mahatma Gandhi*, edited by Shriman Narayan, translated by Madhav Desai. Vol. 1. Ahmedabad: Navajivan Press, 1927.

Holwell, J.Z. *An Account of the Manner of Inoculating for the Small Pox in the East Indies* (London: 1767).

Jones, William. 'Second Anniversary Discourse', *Asiatick Researches*, 1 (1783–1784).

———— 'On the Literature of the Hindus, from Sanscrit, Communicated by Goverdhan Caul, with a Short Commentary', *Asiatick Researches*, Volume 1 (1783–1784).

———— 'The Design of a Treatise of Plants of India', *Asiatick Researches*, 2 (1784), 345–352.

———— 'Ninth Annual Discourse on Asiatick History, Civil and Natural', *Asiatick Researches*, 4 (1794–1795), 1–17.

Legge, James, *A Record of Buddhistic Kingdoms Being an Account by the Chinese Monk Fa?-Hien of his Travels in India and Ceylon (A.D. 399–414) in Search of the Buddhist Books of Discipline* (Oxford: 1886).

Shoobred, John, *Report on the Introduction and Progress of Vaccine Inoculation in Bengal* (Calcutta: 1804).

Shore, John, 'Discourse the Eleventh on the Philosophy of the Asiatics', *Asiatick Researches*, 4 (1794–1795).

Sonnerat, *Voyages to the East Indies and China*, 2.

Tavernier, J.B., *Tavernier's Travels in India* (Calcutta: 1905).

Waring, Edward John, *Bazaar Medicines and Common Medical Plants of India* (London: 1897).
White, Ainslie, *Materia Medica of Hindoostan and Artisan's and Argiculturist's Nomenclature* (Madras: 1813).
Wise, Thomas A., *Commentary on the Hindu System of Medicine* (London: 1845).

## Hindi

Arzani, Muhammad Akbar. *Chikitsa-chakravarti* (Moradabad: 1912).
Chakraborty. *Garbharaksha-a* (Calcutta: 1898).
Devi, Yashoda. *Sacchi Patiprem* (Allahabad: 1910).
———— *Nari Niti Siksha* (Allahabad: 1910).
———— *Grihini Kartavya Sastra* (Allahabad: 1915–1916).
———— *Arogya-Vidhana-grihini-kartavya Sastra* (Allahabad: 1924).
———— *Prana-ballabha Purushvata Vikasa* (Allahabad: 1924).
———— *Dampati Arogyata Jivanshastra artha Ratishastra* (Allahabad: 1924–1925).
———— *Adarsh Pati-Patni Sur Santati-Sudhara* (Allahabad: 1924–1925).
———— *Kanya-Karttavya* (Allahabad: 1925).
———— *Pati Prema Patrika* (Allahabad: 1925).
———— *Nari-Svasthya raksha* (Allahabad: 1925–1926).
———— *Patibrata-Dharmasastra* (Allahabad: 1926–1927).
———— *Dampati Arogyata* (Allahabad: 1927).
———— *Ratnamala* (Allahabad: 1927).
———— *Dampati Arogyata Jivanshastra* (Allahabad: 1931).
———— *Nari-Dharmasastra aur Griha Prabandha Siksha* (Allahabad: 1931).
———— *Swami Dayananda Sarasvati aur Bharata ki Striyon* (Allahabad: 1931).
———— *Vaidyaka Ratna Sangraha* (Allahabad: 1931).
———— *Adarsa Balika-Bhai Bahina* (Allahabad: 1932).
———— *Brahma* (1933).
———— *Prem* (1933).
———— *Dampati Prema aur Ratikriya Guptarahasya* (Allahabad: 1935).
Gandhi, Mohandas Karamchand. *Arogya Sadhana* (Calcutta: 1922).
Gune, Gangadharasastri. *Ayurvediya aushadhi gunadharmasastra* (Ahmednagar: 1935).
Gupta. *Pak-Prakash athava Mithai* (Pratapgarh: 1930).
Gurunncharay. *Hindu Mata athart hindu shishu svastya raksha (Hindu streeyon ke liye shishu svastya aur shishu sambandhi pustak)* (Lahore: 1926).
Premchand. 'Mahila-sabhaon mein santan-nigrah ka prastav', in *Vividh Prasang*, ed. by Premchand (Allahabad: 1932).
Premi, Visvambharasahaya. *Arogya Digdarsana* (Meerut: 1925).
Raichaudhuri, Shaukat. *Ayurvedakavaijnanika swarupa* (Kangri: 1918).
Saheb, Hakim Maulvia Muhammad Abudllah. *Bargad ke Goon Tatha Uske Upyog* (Doheta Pustak Bazaar, 1944).
———— *Anar ke Goon Tatha Uske Upyog* (Doheta Pustak Bazaar, 1944).
———— *Aam ke Goon Tatha Uske Upyog* (Doheta Pustak Bazaar, 1944).
———— *Doodh ke Goon Tatha Uske Upyog* (Doheta Pustak Bazaar, 1944).
Sampadak, Srinarmdeshvar Sharm Bhutpurv *Shisu-Palan* (Calcutta: 1933).
Saraswati, Swami Ramanand, *Chikitsa Darpan*, 2nd edn (Aligarh: 1931).
Sarma, *Paka-Vijnana* (Benares: 1933).

Sarma, Jagganath. *Arogyadarpana* (Allahabad: 1898).

Sarma, Ramprasad. *Ayurveda sutram* (Lahore: 1924–1925).

Sastri, Chatursena. *Arogya sastra* (Delhi: 1932).

Sastri, Saligrama. *Ayurveda-mahatva* (Lucknow: 1925).

Sen, Gananath. *Pratyaksha Shariram* (Dehradun: 1929).

Sharma, Jagannath, *Arogya Darpan* (Prayag: 1915).

Sharma, Ramnarayan. *Dugdh Chikitsa* (Banaras: 1918).

Sharma, Shiv. *Carakasamhita* (Bombay: 1931).

Sharma, Shyamsundar. *Ayurved-Mahantva* (Lucknow: 1925).

Sharma, Srinarmdeshvar. *Shishu Palan* (Banaras: 1933).

Shastri, Pandit Bhairav Prasad Shunk. *Kashi HIndu Vishwavidyalay maom-pyau ko anaubaut yaaoga* (Jagdishpur (Gorakhpur): 1946).

Shastri, Pt. Shalagramji. *Ayurveda-mahattva* (Lucknow: 1925).

Shastri, Shrishalagraam. *Vedon meinTridhatuvad* (Benares: 1922).

Sinha, Dalajit and Ramjit Sinha, *Ayurvediya-kosha* (Etawah: 1932).

Sinha, Ramjit and Daljit Sinha. *Ayurvediya-Kosha/An Encyclopoedical Ayurvedic Dictionary* (Baralokpur-Ithava (United Provinces): 1932).

Sinha, Ramjit and Dalajit Sinha. *Ayurvediya-kosha: an encyclopoedical Ayurvedic dictionary (with full details of Ayurveda, unani and allopathic terms)* (Baralokpur: 1934–1943).

Upadhyaya, Nandalalasarma. *Ayurveda ka mula granthaarthat Brahma-samhita*, ed. by Jodhpur (1938).

Vaid, Haridas. *Chikitsa Chandrodaya* (Allahabad: 1920).

Vaid, Vachspati Shrigalrajsharma. *Srichangani abhinandan granth* (1947).

Vaidya, Haridayal. *Arogya sastra* (Lahore: 1922).

Vaidya, Jagannathaprasad Sukla. *Arogyavidhana athava bharat-men* (Allahabad: 1921–1922).

Varma, Pravasilal. *Arogya mandira* (Benares: 1928).

## Secondary Source Material

Agamben, G., and D. Heller-Roazen. *Homo Sacer: Sovereign Power and Bare Life*. Stanford: Stanford University Press, 1998.

Ahluwalia, Sanjam. *Reproductive Restraints: Birth Control in India, 1877–1947*. University of Illinois Press, 2007.

Alavi, Seema. *Islam and Healing: Loss and Recovery of an Indo-Muslim Medical Tradition 1600–1900*. Delhi: Permanent Black, 2007.

——— "Medical Culture in Transition: Mughal Gentleman Physician and the Native Doctor in Early Colonial India." *Modern Asian Studies* 42, no. 5 (September 2008): 853–897.

——— "Unani Medicine in the Nineteenth-century Public Sphere: Urdu Texts and the Oudh Akhbar." *Indian Economic and Social History Review* 42, no. 1 (2005): 102–129.

Alter, J. *Gandhi's Body: Sex, Diet and the Politics of Nationalism*. Philadelphia: University of Pennsylvania Press, 2000.

——— *The Wrestler's Body: Identity and Ideology in North India*. Berkeley: University of California Press, 1992.

Alter, J., F. Bray, A. Guha, P.C. Joshi, and C. Leslie. "Heaps of Health, Metaphysical Fitness: Ayurveda and the Onthology of Good Health in Medical Anthropology (and Comments and Reply)." *Current Anthropology* 40, no. Supplement: Special Issue: Culture, A Second Chance? (1999): S43–S66.

Amrith, S.S. *Decolonizing International Health: India and Southeast Asia, 1930–1965.* Basingstoke: Palgrave Macmillan, 2006.

—— "Food and Welfare in India, C. 1900–1950." *Comparative Studies in Society and History* 50, no. 4 (2008): 1010–1035.

Anagol, P. "The Emergence of the Female Criminal in India: Infanticide and Survival Under the Raj." *Indian Economic and Social History Review* 53, no. 1 (Spring 2002).

Anderson, Warwick. "Where Is the Postcolonial History of Medicine?" *Bulletin of the History of Medicine* 72, no. 3 (1998): 522–530.

—— *Colonial Pathologies: American tropical medicine, race, and hygiene in the Philippines.* Duke University Press Books, 2006.

Antia, N.H. "An Alternative Strategy for Health Care? The Mandwa Project." *Economic and Political Weekly* 20, no. 51/52, 21 December 1985: 2257–2260.

Appadurai, A. "Numbers in the Colonial Imagination." In *Orientalism and the Postcolonial Predicament: Perspectives on South Asia,* edited by P. Van de Veer and C. Breckenridge. Philadelphia: University of Pennsylvania Press, 1993.

Arnold, D. *Colonizing the Body: State Medicine and Epidemic Disease in Nineteenth-Century India.* Berkeley: University of California Press, 1993.

—— "Official Attitudes to Population, Birth Control and Reproductive Health in India, 1921–1946." In *Reproductive Health in India: History, Politics, Controversies,* edited by S.E. Hodges. Delhi, 2006.

Arnold, D. "Plurality and Transition: Knowledge Systems in Nineteenth Century India." presented at the Princeton History of Science Seminar, Princeton, October 2003.

—— "The Armed Police and Colonial Rule in South India, 1914–1947." *Modern Asian Studies* 11, no. 1 (1977): 101–125.

Attewell, Guy. *Refiguring Unani Tibb.* Hyderabad: Orient Longman, 2007.

Bagchi, J. "Representing Nationalism: Ideology of Motherhood in Colonial Bengal." *Economic and Political Weekly,* 20 October 1990.

Bala, P. *Imperialism and Medicine in Bengal: A Socio-historical Perspective.* New Delhi: Sage, 1991.

Basham, A.L. *A Cultural History of India.* Oxford: OUP, 1975.

Bashford, Alison. *Imperial Hygiene: A Critical History of Colonialism, Nationalism and Public Health.* Basingstoke: Palgrave Macmillan, 2004.

Bayly, C.A. *Empire and Information: Intelligence Gathering and Social Communication in India, 1780–1870.* Cambridge: CUP, 1996.

—— "Rammohan Roy and the Advent of Constitutional Liberalism in India, 1800–30." *Modern Intellectual History* 4, no. 01 (2007): 25–41.

—— *Recovering Liberties: Indian Thought in the Age of Liberalism and Empire.* Cambridge: Cambridge University Press, 2012.

Benner, D. "Traditional Indian Systems of Healing and Medicine: Ayurveda." In *Encyclopedia of Religion,* 2004.

Berger, Rachel. "Between Digestion and Desire: Genealogies of Food in Nationalist North India." *Modern Asian Studies FirstView* (2013): 1–22.

———— "From the Biomoral to the Biopolitical: Ayurveda's Political Histories." *South Asian History and Culture* 4, no. 1 (2013): 48–64.

———— "Ayurveda and the Making of the Urban Middle Class in North India, 1900–1945." In Dagmar Wujastyk, and Frederick M. Smith (eds). *Modern and Global Ayurveda: Pluralism and Paradigms*, 101–15. Buffalo: SUNY Press, 2008.

Bhattacharya, S. *Expunging Variola: The Control and Eradication of Smallpox in India, 1947–1977.* London: Sangam, 2006.

Bose, S. *His Majesty's Opponent: Subhas Chandra Bose and India's Struggle against Empire.* Cambridge: Harvard University Press, 2011.

Bose, S., and Kris Manjapra, eds. *Cosmopolitan Thought Zones: South Asia and the Global Circulation of Ideas.* London: Palgrave Macmillan, 2010.

Brass, P.R. "A Reply to Francis Robinson." *Journal of Commonwealth and Comparative Politics* XV (1977): 231–233.

———— *Language, Religion and Politics in North India.* Cambridge: CUP, 1974.

Brennan, L. "The Illusion of Security: The Background to Muslim Separatism in the United Provinces." *Modern Asian Studies* 18, no. 2 (1984): 237–272.

Burchell, G., C. Gordon, and P. Miller. *The Foucault Effect: Studies in Governmentality.* London: Harvester Wheatsheaf, 1991.

Burton, A. "Contesting the Zenana: The Mission to Make 'Lady Doctors for India,' 1874–1885." *The Journal of British Studies* 35, no. 3 (1996): 368–397.

Carroll, L. "Law, Custom, and Statutory Social Reform: The Hindu Widow's Remarriage Act of 1856." In *Women in Colonial India: Essays on Survival, Work, and State*, edited by J. Krishnamurthi, 1–26. Oxford: Oxford University Press, 1989.

Chakrabarty, D. *Habitations of Modernity: Essays in the Wake of Subaltern Studies.* University of Chicago Press, 2002.

Chakravarti, U. "Whatever Happened to the Vedic Dasi?: Orientalism, Nationalism and a Script for the Past." In *Recasting Women*, edited by K. Sangari and S. Vaid. New Delhi: Kali for Women, 1989.

Chancellor, N. "A Picture of Health: The Dilemma of Gender and Status in the Iconography of Empire, India C. 1805." *Modern Asian Studies* 35, no. 4 (2001): 769–782.

Chatterjee, P. "The Nationalist Resolution of the Women's Question." In *Recasting Women: Essays in Colonial History*, edited by K. Sangari and S. Vaid. Delhi: Kali for Women Press, 1989.

Chatterjee, Partha. *The Nation and Its Fragments: Colonial and Postcolonial Histories.* Princeton: Princeton University Press, 1993.

Chattopadhyaya, D.P. *History of Science and Technology in Ancient India.* Calcutta: Firma KLM, 1986.

Chowdhury-Sengupta, I. "Mother India and Mother Victoria: Motherhood and Nationalism in Nineteenth-Century Bengal." *South Asia Research* 1, no. May (1992): 20–37.

Cohen, Lawrence. "The Other Kidney: Biopolitics Beyond Recognition." *Body & Society* 7, no. 2–3 (2001): 9 –29.

Collingham, E.M. *Curry: A Tale of Cooks and Conquerors.* London: Vintage Books, 2005.

———— *Imperial Bodies: The Physical Experience of the Raj, 1800–1947.* Cambridge: Polity Press, 2001.

Copeman, Jacob. *Veins of Devotion: Blood Donation and Religious Experience in North India*. Camden, NJ: Rutgers University Press, 2009.

Daechsel, M. *The Politics of Self-Expression: The Urdu Middle-class Milieu in Mid-twentieth Century India and Pakistan*. Abingdon: Routledge, 2006.

Dalmia, V. *The Nationalization of Hindu Traditions: Bharatendu Harishchandra and Nineteenth Century Benaras*. Delhi: Oxford University Press, 1997.

Datta, Manmathnatha. *Ayurveda; or the Hindu System of Medical Science*. Calcutta: H. C. Dass, 1899.

Davis, Mike. *Late Victorian Holocausts: El Niño Famines and the Making of the Third World*. Verso, 2002.

Deb Roy, Rohan. "Debility, Diet, Desire: Food in Nineteenth and Early Twentieth Century Bengali Manuals." In *The Writer's Feast: Food and the Politics of Representation*, edited by Surpriya Chaudhuri and Rimi B Chatterjee, 179–205. Hyderabad: Orient Blackswan, 2011.

Dirks, N. B. *Castes of Mind: Colonialism and the Making of Modern India*. Princeton: Princeton University Press, 2001.

Dodson, Michael S. *Orientalism, Empire, and National Culture: India, 1770–1880*. Basingstoke: Palgrave Macmillan, 2007.

Doniger, W., and S. Kakar. *Vatsyayana's Kama Sutra*. Oxford: OUP, 2002.

Ernst, W. *Mad tales from the Raj: the European insane in British India, 1800–1858*. London: Routledge, 1991.

——— "Beyond East and West: From the History of Colonial Medicine to a Social History of Medicine (s) in South Asia." *Social History of Medicine* 20, no. 3 (2007): 505–524.

Fazl, Abu. *The Ain-iAkbari by Abul Fazl-i-Allami*. Translated by H.S. Jarret. Calcutta: Asiatic Society of Bengal, 1894.

Forbes, G. "Managing Midwifery in India." In *Contesting Colonial Hegemonies: State and Society in Africa and India*, edited by D. Engels and S. Marks. London: British Academy Press, 1994.

Foucault, M. "The Confession of the Flesh." *Power/knowledge: Selected Interviews and Other Writings* 1977 (1972): 194–228.

——— *The History of Sexuality: Vol 1, The Will to Knowledge*. Harmondsworth: Penguin, 1990.

——— *The Birth of Biopolitics: Lectures at the College De France 1978–1979*. New York: Picador, 2010.

Fraser, N. "What's Critical About Critical Theory? The Case of Habermas and Gender." *New German Critique* (1985): 97–131.

Fraser, N., and C. Calhoun, eds. *Habermas and the Public Sphere*. Cambridge: MIT Press, 1992.

Gallagher, J., G. Johnson, and A. Seal. *Locality, Province and Nation: Essays on Indian Politics, 1870 to 1940*. London: CUP, 1973.

Goswami, M. *Producing India: From Colonial Economy to National Space*. Chicago: University of Chicago Press, 2004.

Gould, W. *Hindu Nationalism and the Language of Politics in Late Colonial India*. Cambridge: Cambridge University Press, 2004.

Gupta, C. "Procreation and Pleasure Writings of a Woman Ayurvedic Practitioner in Colonial North India" in *Studies in History* 21, 1 (February 2005): 17–44.

——— *Sexuality, Obscenity, Community: Women, Muslims, and the Hindu Public in Colonial India*. Delhi: Permanent Black, 2002.

Habermas, J. *Structural Transformation of the Public Sphere: An Inquiry into a Category of Bourgeois Society*. Cambridge: MIT Press, 1991.

Habib, I., and D. Raina. *Domesticating Modern Science: Essays on Social History of Science and Culture in Colonial India*. New Delhi: Tulika, 2004.

Hancock, M. "Home Science and the Nationalization of Domesticity in Colonial India." *Modern Asian Studies* 35, no. 4 (2001): 871–903.

Hansen, Thomas Blom. *The saffron wave: Democracy and Hindu nationalism in modern India*. Princeton: Princeton University Press, 1999.

Haraway, D.J. *Simians, Cyborgs, and Women: The Reinvention of Nature*. 41. Routledge, 1991.

Hardiman, David. "Indian Medical Indigeneity: From Nationalist Assertion to the Global Market." *Social History* 34, no. 3 (2009): 263–283.

Harrison, M. *Public Health in British India: Anglo-Indian Preventive Medicine 1859–1914*. Cambridge: Cambridge University Press, 1994.

Haynes, D., and S. Botre. "R.D. Karve and Male Middle-Class Sexuality in Western India: Correspondence Published in Samaj Swasthya, 1927–53," forthcoming.

Hodges, S.E. *Contraception, Colonialism and Commerce: Birth Control in South India, 1920–1940*. Aldershot: Ashgate, 2008.

——— "Governmentality, Population and Reproductive Family in Modern India." *Economic and Political Weekly* (2004): 1157–1163.

——— "Looting the Lock Hospital in Colonial Madras During the Famine Years of the 1870s." *Social History of Medicine* 18, no. 3 (2005): 379–398.

——— *Reproductive Health in India: History, Politics, Controversies*. Vol. 13. Hyderabad: Orient Longman, 2006. http://wrap.warwick.ac.uk/38898/

Joshi, S. *Fractured Modernity: Making of a Middle Class in Colonial North India*. Delhi: Oxford University Press, 2002.

Kapila, Shruti. *An Intellectual History for India*. Cambridge: Cambridge University Press, 2011.

Karambelkar, V.W. *The Atharva-Veda and the Ayur-Veda*. Nagpur, 1961.

Kasturi, M. "Law and Crime in India: British Policy and the Female Infanticide Act of 1870." *Indian Journal of Gender Studies* 1, no. 2 (1994): 169–194.

Kaviraj, S. *The Unhappy Consciousness: Bankimchandra Chattopadhyay and the Formation of Nationalist Discourse in India*. Delhi: Oxford University Press, 1995.

Khan, Yasmin. *The Great Partition: The Making of India and Pakistan*. New Haven: Yale University Press, 2007.

King, C.R. *One Language, Two Scripts: The Hindi Movement in Nineteenth Century India*. Bombay: Oxford University Press, 1994.

Kishwar, Madhu. *Gandhi and Women*. Delhi: Manushi Prakashan, 1986.

Kumar, A. *Medicine and the Raj: British Medical Policy in India, 1835–1911*. New Delhi: Sage, 1998.

Lal, M. " 'The Ignorance of Women Is the House of Illness': Gender, Nationalism, and Health Reform in Colonial North India." In *Medicine and Colonialism: The Politics of Identity*, edited by M.P. Sutphen. London, 2003.

——— "The Politics of Gender and Medicine in Colonial India: The Countess of Dufferin Fund, 1885–1888." *Bulletin of the History of Medicine* 68 (1994): 29–66.

Lal, V. "Nakedness, Non-Violence and the Negation of Negation: Gandhi's Experiments in Brahmachary and Celibate Sexuality." *South Asia* 22, no. 2 (1999): 63–94.

——— "Subaltern Studies and Its Critics: Debates over Indian History." *History and Theory* 40, no. 1 (2001): 135–148.

Lang, S. "Drop the Demon Dai: Maternal Mortality and the State in Madras Presidency 1840–1875" (2003).

Langford, Jean. *Fluent Bodies: Ayurvedic Remedies for Postcolonial Imbalance*. Durham: Duke University Press, 2002.

Legg, Stephen. "Of Scales, Networks and Assemblages: The League of Nations Apparatus and the Scalar Sovereignty of the Government of India." *Transactions of the Institute of British Geographers* 34, no. 2 (2009): 234–253.

——— "Planning Social Hygiene: From Contamination to Contagion in Interwar India." In *Imperial Contagions: Medicine, Hygiene, and Cultures of Planning in Asia*, edited by Robert Peckham, 105–122. Hong Kong: Hong Kong University Press, 2013.

——— *Spaces of Colonialism: Delhi's Urban Governmentalities*. Oxford: Wiley-Blackwell, 2007.

——— "Stimulation, Segregation and Scandal: Geographies of Prostitution Regulation in British India, Between Registration (1888) and Suppression (1923)." *Modern Asian Studies* 46, no. 06 (2012): 1459–1505.

——— "Transnationalism and the Scalar Politics of Imperialism." *New Global Studies* 4, no. 1 (2010): 1–7.

Lelyveld, D. "Colonial Knowledge and the Fate of Hindustani." *Comparative Studies in Society and History* 35, no. 4 (1993): 665–682.

Leslie, Charles A. *Asian Medical Systems*, Berkeley: University of California, 1976.

Levine, P. *Prostitution, Race and Politics: Policing Venereal Disease in the British Empire*. London: Routledge, 2003.

Louro, Michele L. "Rethinking Nehru's Internationalism: The League Against Imperialism and Anti-imperial Networks, 1927–1936." *Third Frame: Literature, Culture and Society* 2, no. 3 (September 2009): 79–94.

Mani, L. "Contentious Traditions: The Debate on Sati in Colonial India." In *Recasting Women: Essays in Colonial History*, edited by K. Sangari and S. Vaid, 88–126. Kali for Women, 1989.

Marks, S. "What Is Colonial About Colonial Medicine? And What Has Happened to Imperialism and Health?" *Social History of Medicine* 10, no. 02 (1997): 205–219.

Marriott, M. *India Through Hindu Categories*. Delhi: Sage, 1990.

——— "Interpreting Indian Society: A Monistic Alternative to Dumont's Dualism." *The Journal of Asian Studies* 36, no. 1 (1976): 189–195.

Marriott, M., and R. Inden. "Towards an Ethnosociology of South Asian Caste Systems." In *The New Wind: Changing Identities in South Asia*, edited by K. David. The Hague: Mouton Publishers, 1977.

McGinn, P.A. "The Age of Consent Act (1891) Reconsidered: Women's Perspectives and Participation in the Child Marriage Controversy in India." *South Asia Research* 12, no. 2 (1992): 100–118.

Mukharji, P.B. "Enframing Bangali Ayurbed: Going Beyond 'Frontier' Frames." *Journal of the Asiatic Society of Bangladesh (Humanities)* 49, no. 1 (June 2004): 13–40.

——— *Nationalizing the Body: The Medical Market, Print and Daktari Medicine*. London: Anthem Press, 2009.

Nicholas, R.W. "The Godess Sitala and Epidemic Smallpox in Bengal." *Journal of Asian Studies* 41, no. 1 (1981): 21–44.

O'Hanlon, Rosalind. "Recovering the Subject: Subaltern Studies and Histories of Resistance in Colonial South Asia." *Modern Asian Studies* 22, no. 1 (1988): 189–224.

Orsini, F. *The Hindi Public Sphere, 1920–1940: Language and Literature in the Age of Nationalism*. Delhi: Oxford University Press, 2002.

Pande, Ishita. *Medicine, Race and Liberalism in British Bengal: Symptoms of Empire*. London: Routledge, 2009.

Pandey, G. *The Ascendancy of the Congress in Uttar Pradesh: Class, Community and Nation in Northern India, 1920–1940*. London: Anthem, 2002.

Pandey, Gyanendra. *The Construction of Communalism in Colonial North India*. Delhi: Oxford University Press, 2006.

Parry, J. *Death in Banaras*. Cambridge: Cambridge University Press, 1994.

Pati, B. "Negotiating with Dharma Pinnu: Towards a Social History of Smallpox in Colonial Orissa." *Canadian Bulletin Medical History* 19, no. 2 (2002): 477–492.

Peers, D.M. "Privates Off Parade: Regiments and Sexuality in the Nineteenth Century Indian Empire." *International History Review* 20, no. 4 (1998): 823–854.

Philip, Kavita. *Civilizing Natures: Race, Resources, and Modernity in Colonial South India*. Camden: Rutgers University Press, 2003.

Pollock, Sheldon. *The Language of the Gods in the World of Men: Sanskrit, Culture, and Power in Premodern India*. Berkeley: University of California Press, 2006.

Prakash, G. *Another Reason: Science and the Imagination of Modern India*. Princeton: Princeton University Press, 1999.

——— "Subaltern Studies as Postcolonial Criticism." *The American Historical Review* 99, no. 5 (1994): 1475–1490.

Premchand. "Mahila-sabhaon Mein Santan-nigrah Ka Prastav'." In *Vividh Prasang*, edited by Premchand, 252. Allahabad, 1932.

Rai, A. *Hindi Nationalism*. Delhi: Three Essays Collective, 2001.

Ramasubban, R. "Imperial Health in British India 1857–1900." In *Disease, Medicine & Empire: Perspectives on the Medicine and the Experience of European Expansion*, edited by R.M. Macleod and M. Lewis. London: Routledge, 1988.

——— *Public Health and Medical Research in India*. Stockholm: Swedish Agency for Research Cooperation with Developing Countries, 1982.

Ramusack, B. "Embattled Advocates: The Debate Over Birth Control in India,1920–1940." *Journal of Women's History* 1, no. 2 (1989): 34–64.

Razzack. *Hakim Ajmal Khan, the Versatile Genius*. New Delhi: Central Council for Research in Unani Medicine, Ministry of Health and Family Welfare, 1987.

Renold, L. *A Hindu Education: Early Years of Banaras Hindu University*. New Delhi: OUP, 2005.

Robinson, F. " 'Nation Formation: The Brass Thesis and Muslim Separatism'." *Journal of Commonwealth and Comparatiue Politics* XV (1977): 215–30.

Scott, J.W. *Gender and the Politics of History*. New York: Columbia University Press, 1989.

Seal, A. *The Emergence of Indian Nationalism: Competition and Collaboration in the Later Nineteenth Century*. Vol. 1. Cambridge: Cambridge University Press, 1971.

Sen, Samita. "Motherhood and Mothercraft: Gender and Nationalism in Bengal." *Gender & History* 5, no. 2 (1993): 231–243.

Sharma, Madhuri. *Indigenous and Western Medicine in Colonial India*. Delhi: Cambridge University Press, 2012.

Sherman, T. C. *State Violence and Punishment in India*. London: Routledge, 2009.

Shohat, E. "Notes on the 'Post-Colonial'." *Social Text* no. 31/32 (1992): 99–113.

Sinha, M. "The Lineage of the 'Indian' Modern: Rhetoric, Agency and the Sarda Act in Late Colonial India." In *Gender, Sexuality and Colonial Masculinities*, edited by A. Burton, 207–221. New York, 1999.

―――― *Specters of Mother India: The Global Restructuring of an Empire*. Durham: Duke University Press, 2006.

Sinha, R. and D. Sinha. *Ayurvediya-kosha: An Encyclopoedical Ayurvedic Dictionary (with Full Details of Ayurveda, Unani and Allopathicterms)*. Baralokpur, 1934.

Sivaramakrishnan, Kavita. *Old Potions, New Bottles: Recasting Indigenous Medicine in Colonial Punjab*. Hyderabad: Orient Longman, 2006.

Stoler, Ann Laura. *Race and the Education of Desire: Foucault's History of Sexuality and the Colonial Order of Things*. Durham: Duke University Press, 1995.

Stolte, Carolien. " 'Enough of the Great Napoleons!' Raja Mahendra Pratap's Pan-Asian Projects (1929–1939)." *Modern Asian Studies* 46, no. Special Issue 02 (2012): 403–423.

Tarlo, E. *Clothing Matters: Dress & Its Symbolism in Modern India*. Chicago: University of Chicago Press, 1996.

Tejani, Shabnum. Tejani, Shabnum. *Indian secularism: A social and intellectual history, 1890–1950*. Bloomington: Indiana University Press, 2008

Thapar, R. *Early India: From the Origin to AD 1300*. London: Penguin, 2002.

―――― "Renunciation: The Making of a Counter-culture?" In *Cultural Pasts: Essays in Early Indian History*, edited by R. Thapar, 876–913. Delhi: Oxford University Press, 2000.

―――― *The Aryan: Recasting Constructs*. Gurgaon: Three Essays Collective, 2008.

Tomlinson, B.R. *The Political Economy of the Raj, 1914–1947: The Economics of Decolonization in India*. London: Macmillan, 1979.

Torri, M. " 'Westernized Middle-Class', Intellectuals and Society in Late Colonial India." In *The Congress and Indian Nationalism: Historical Perspectives*, edited by J.L. Hill. London: Curzon, 1991.

Travers, R. "The Eighteenth Century in Indian History: A Review Essay." *Eighteenth-Century Studies* 40, no. 3 (2007): 492–508.

Trivedi, L.N. "Visually Mapping the 'Nation': Swadeshi Politics in Nationalist India, 1920–1930." *The Journal of Asian Studies* 62, no. 1 (February 2003): 11–41.

Varma, D. "From Witchcraft to Allopathy: Uninterrupted Journey of Medical Science." *Economic and Political Weekly* (February 19, 2006): 3605–3611.

―――― *Reason and Medicine: Art and Science of Healing from Antiquity to Modern Times*. Gurgaon: Three Essays Collective, 2012.

Vaughan, M. *Curing Their Ills: Colonial Power and African Illness*. Cambridge: Cambridge University Press, 1991.

Warner, Michael. *Publics and Counterpublics*. New York: Zone Books, 2002.

Washbrook, D.A. "The Rhetoric of Democracy and Development in Late Colonial India." In *Nationalism, Democracy and Development*, edited by S. Bose and A. Jalal. Delhi: OUP, 1997.

Watt, C.A. *Serving the Nation: Cultures of Service, Association, and Citizenship in Colonial India*. Delhi: Oxford University Press, 2005.

White, D. *The Alchemical Body: Siddha Traditions in Medieval India*. Chicago: University of Chicago Press, 1996.

—— *Kiss of the Yogini: "Tantric Sex" in Its South Asian Contexts*. Chicago: Chicago University Press, 2003.

Winter, H.J.J. "Science." In *A Cultural History of India*, edited by A.L. Basham. 2nd ed. Delhi: OUP, 1997.

Wise, Thomas Alexander. *Thomas Alexander, Commentary on the Hindu System of Medicine*. Calcutta: Thacker, 1845.

Wujastyk, Dagmar. *Well-Mannered Medicine: Medical Ethics and Etiquette in Classical Ayurveda*. Oxford: Oxford University Press, 2012.

Wujastyk, Dominik. *The Roots of Ayurveda*. 3rd ed. London: Penguin, 2003.

Young, A. (eds) *Paths to Asian Medical Knowledge*. Berkeley: University of California, 1992.

Zachariah, Benjamin. *Developing India: An Intellectual and Social History c. 1930–50*. New Delhi: Oxford University Press, 2001.

—— Playing the Nation Game: The Ambiguities of Nationalism in India. Calcutta: Yoda Press, 2011.

Zimmermann, F. *The Jungle and the Aroma of Meats*, 1982.

Zysk, K.G. *Ascetism and Healing in Ancient India: Medicine in the Buddhist Monastery*. Oxford: Oxford University Press, 1991.

—— *Medicine in the Veda: Religious Healing in the Veda with Translations from the Rg Veda and the Atharvaveda and Renderings from Corresponding Ritual Texts*. Vol. 3rd. Delhi: Motilal Banarsidass, 1996.

# Index

229